T0258635

IET HISTORY OF TECHNOLOGY SERIES 29

Series Editors: Dr B. Bowers
Dr C. Hempstead

Sir Charles Wheatstone FRS

1802–1875

Other volumes in this series:

Sir Charles Wheatstone FRS

1802–1875

Brian Bowers

The Institution of Engineering and Technology in association with The Science Museum, London

Published by The Institution of Engineering and Technology, London, United Kingdom

First edition © 1975 The Science Museum, London
Second edition © 2001 The Science Museum, London

First published 1975
Second edition 2001
Reprinted with new cover 2012

The Institution of Engineering and Technology
Michael Faraday House
Six Hills Way, Stevenage
Herts, SG1 2AY, United Kingdom

www.theiet.org

British Library Cataloguing in Publication Data
A catalogue record for this product is available from the British Library

ISBN (10 digit) 0 85296 103 0
ISBN (13 digit) 978-0-85296-103-2

Typeset in the UK by RefineCatch Ltd, Bungay, Suffolk
First printed in the UK by MPG Books Ltd, Bodmin, Cornwall
Reprinted in the UK by Lightning Source UK Ltd, Milton Keynes

Artist's impression of the Enchanted Lyre *demonstration, which first brought Wheatstone into prominence.*

Contents

Foreword

The first edition of this book was published in 1975 by HMSO for the Science Museum when I was Director and the author was a member of the curatorial staff. I wrote then:

There are more than one hundred objects made by, designed by or associated with Wheatstone in the Science Museum. Dr Brian Bowers, who succeeded me in having charge of the Electrical Engineering Collection, has been interested in Wheatstone for about ten years. His interest was roused by a study he made of early patent history – while an Examiner at the Patent Office – which included, in particular, Cooke and Wheatstone's first electric telegraph patent. That patent is apparently of as much interest to patent historians as it is to historians of science and technology.

Being a King's College London man, Dr Bowers was already aware of and interested in the considerable influence which Wheatstone exercised there, and from there on the scientific world. So it is a fairly natural consequence that when he found there was a bulk of interesting material about Wheatstone needing to be made known he was moved to examine the surviving records and apparatus, and write this book.

Twenty-six years later Wheatstone remains a major figure in the early history of electrical engineering and the electric telegraph. I am glad that this second edition is being produced and that Brian Bowers has had the opportunity to include more about the concertina, which was the most important of Wheatstone's contributions to musical instruments. The concertina, and its inventor, will shortly be featured in a new Musical Instrument Gallery at the Horniman Museum, where I was the first Chairman of Trustees. That makes it an additional pleasure for me to introduce this revised edition.

Margaret Weston
September 2001

There was nothing he touched that he did not adorn

(Michael Faraday)

Faraday's remark about Wheatstone was quoted by W.H. Preece in a Royal Institution Discourse on Wheatstone in 1880. The expression was first used by Dr Johnson about Oliver Goldsmith. Johnson also said 'there was nothing he did not touch'; there is no record that Faraday said that of Wheatstone, but he might well have done.

Preface

The first edition of this book sold out long ago, but interest in the history of technology is growing and many people have urged me to prepare a new edition. Wheatstone was a key figure in the early development of electrical engineering, and especially the electric telegraph. It was in connection with the telegraph that the Prince Consort wrote to Wheatstone, and I acknowledge the gracious permission of Her Majesty The Queen to publish that letter.

I have been interested in Wheatstone for many years and over that time many people have assisted me by discussing points and helping to find material. It would be difficult, if not impossible, to mention them all by name, but I am grateful to every one. In the first edition I gave special mention to three people: Professor E.J. Burge who, when he was Reader in Physics at King's College London, encouraged me to convert a casual interest in Wheatstone and his telegraph patents into the PhD thesis which was the basis of this book; my wife who has shared this interest in Wheatstone for many years and has typed reams about him, picking up errors and obscurities on the way; and my friend and colleague, the late Keith Geddes, who read every word, commenting on many of them, and made it a better book.

Many of my colleagues in the Science Museum helped me with the history of their various subjects – for Wheatstone's interests were very wide. The Librarians and their staffs at King's College London, the Institution of Electrical Engineers, the Royal Society, the Royal Institution, the Royal Society of Arts and Gloucester City Library gave invaluable assistance. The late Mr H.A. Harvey of King's College Library shared his extensive knowledge of the history of the College. Roy Barker, who has made a study of the electrical history of South Wales, showed me the site of Wheatstone's submarine telegraph trials in Swansea Bay. The late Professor James Greig, who taught me electrical engineering at King's, the late Professor Eric Laithwaite, who was fascinated by unusual electrical machines, and the late Dr E.A. Marland, an historian of the early telegraph, all took a helpful interest in my work.

I first became interested in Wheatstone through my interest in electrical history, but that was not the only field in which he made major contributions. Since the first edition of this book was published much research has been done, especially by Neil Wayne, into the history of the concertina and the Wheatstone musical instrument business. The Concertina Museum Collection, which he gathered, is now in the Horniman Museum in London. I am grateful to Neil for allowing me to draw on some of his work and to Margaret Birley of the Horniman Museum and Gardens for providing concertina pictures.

When writing the first edition I was not aware of Wheatstone's wave machine, an ingenious device designed to help in understanding wave theory. I gladly acknowledge the assistance of three people who helped me to understand the working of the machine: my former colleague, Mr V.K. Chew, who first drew my attention to Professor Secchi's 1849 paper in Italian which carries the earliest description of Wheatstone's wave machine; Julian Holland of the Macleay Museum of the University of Sydney, New South Wales, who has made a study of the sixteen known examples; and Paolo Brenni of the Instituto e Museo di Storia della Scienza in Florence who provided me with a photograph of the wave machine in the University of Pavia.

The first edition was prepared on a typewriter, and then published by HMSO for the Science Museum. The first stage in preparing this revised and slightly enlarged second edition, which is being published by the Institution of Electrical Engineers in association with the Science Museum, was to 'scan' the original text into a computer. I am grateful to my sister, Helen Purves, for doing much of this, using modern computer technology which would have delighted Wheatstone. The first edition was originally Crown Copyright but the rights have been transferred to the Science Museum. I am grateful to Ela Ginalska, the Science Museum Publications Manager, for permitting this new edition and for making arrangements with the Museum's Science and Society Picture Library for photographs from the Museum's collections. All the photographs in this edition, except those specifically credited to others, were supplied by the Science and Society Picture Library. I am grateful also to the IEE Publications Department, especially Roland Harwood and Diana Levy, for seeing this book through to publication, and to Dame Margaret Weston for contributing the Foreword.

Brian Bowers
August 2001

List of figures

List of tables

Chapter 1

An extraordinary fellow

Charles Wheatstone received a caller at his home in Conduit Street, London, on 27 February 1837. The caller, W.F. Cooke, was trying to develop an electric telegraph and had been told that Professor Wheatstone might be able to help him with some scientific difficulties that had arisen. Cooke had already consulted several eminent scientists without success, but after this meeting he wrote enthusiastically 'Wheatstone is the only man near the mark . . . an extraordinary fellow'.[1]

Within a few years Wheatstone was to become a household name for his work on electric telegraphs, and indeed the Prince Consort consulted him as a parent of the telegraph system. Wheatstone has other claims to fame. He held a Chair at King's College London for 41 years and, although he hardly ever gave a lecture, the College subsequently named a laboratory after him. He invented the concertina and he discovered the principle of stereoscopy. He used his encyclopaedic knowledge of the literature to spread scientific ideas. He designed ingenious electro-mechanical devices and pioneered precise electrical measurements.

Ohm's Law and Wheatstone's Bridge were introduced in Wheatstone's 1843 paper on electrical measurements. Both had been published previously but less prominently, and Wheatstone clearly acknowledged the source of each. The bridge was devised by S.H. Christie, yet it has been known ever since as Wheatstone's. It is purely by chance that we speak now of 'Ohm's Law' not 'Wheatstone's Law'.

Most of this book is based on Wheatstone's scientific publications and the records of his scientific and business activities. There is very little record of his private life, though a few glimpses may be gleaned from the diaries, correspondence and reminiscences of people with whom he came in contact.

The portraits and photographs mostly show him in later life and give the impression of an astute yet kindly gentleman. In marked contrast is the drawing (Figure 7.1) by William Brockendon, the only dated portrait (20 November 1837), which shows a curly-headed, youthful figure more reminiscent of one of the romantic poets than a rising Professor at King's College. In appearance he closely resembled his friend Lyon Playfair, whom the *Dictionary of National Biography* describes as being 'below average height and strikingly intellectual in appearance'. They probably first met at British Association meetings. Playfair wrote in his *Memoirs* that Wheatstone:

though older, was, like myself, small of stature, and we both wore spectacles. We were constantly mistaken for each other, and we must have been alike, for once Lady Wheatstone addressed me as her husband.[2]

(Wheatstone's wife was never 'Lady Wheatstone', for she died before her husband was knighted.) Another friend, Leopold Martin, gave a somewhat more revealing sketch of Wheatstone in his *Reminiscences*:

Sir Charles Wheatstone was small in feature, childlike to a degree, short-sighted and with wonderful rapid utterance, yet seemingly quite unable to keep pace with an overflowing mind.[3]

Behind the sombre figure of the man of science totally engrossed in his work, we have glimpses of a keen sense of fun, a delight in harmless pranks and a man who could be the life and soul of a private party, although painfully shy in public. Wheatstone and Playfair used to amuse themselves by deciphering the cipher advertisements in the personal column of *The Times*.

In September 1846 Wheatstone was one of a party staying at the home of Florence Nightingale's father. Sir Roderick Murchison, a fellow guest, recorded in his journal that:

Wheatstone was of the party, and he engaged to perform the trick of the invisible girl, by telling you what was in places where no one could see anything. But to do this a confeder-ate was required, and peering into the faces of all the women, he selected Florence as his accomplice, and, having taken her out of the room for half an hour, they came back and performed the trick. On talking to my friend about the talent of the girl, he said, 'Oh! if I had no other means of living I could go about to fairs with her and pick up a deal of money'.[4]

Neither Murchison nor his biographer explained the 'trick of the invisible girl'. Florence Nightingale (1820–1910) was as yet an unknown young woman. In 1853, at the start of the Crimean War, she was invited to start a training school for nurses at King's College Hospital, and the rest is history!

Wheatstone had already made a name for himself before embarking on the telegraph and before meeting Cooke. He was born into a family of musical instrument makers, music publishers and music teachers, but he was not content merely to make and sell traditional instruments. From his youth he conducted experiments on the instruments he made and he studied the scientific principles involved. It was this that brought him to the notice of the scientific establishment and led him to a career in science.

Wheatstone's interest in telegraphy arose from his study of sound. He began with a study of the transmission of sound through rods and stretched wires, investigating in particular how far it was possible to transmit sound through solid bodies, before he turned to electric telegraphy. The possible analogy between sound and light led Wheatstone to a study of vision and an understanding of the principle of binocular vision. The outcome was the stereoscope which, improved by Brewster and with the aid of photography, became one of the most popular scientific toys of the nineteenth century.

Wheatstone's work on the electric telegraph had a greater impact on daily life than anything else he did. His partnership with Cooke was short-lived, but it is central to our story. The two men should have formed an ideal partnership. Wheatstone had the scientific understanding and technical skill to make a working telegraph, though initially he looked on his work as a piece of pure research which might have a useful application. Cooke was an entrepreneur who could see the financial rewards a successful telegraph would bring, and he had the initiative and enthusiasm to establish a business. Alas, they quarrelled and the partnership degenerated into a series of bitter disputes which marred both their lives. Nevertheless, the telegraph they pioneered brought them fame, honour and fortune, and brought the world a revolution in communications.

Notes

1 Letter, Cooke to his mother, 27 February 1837, in the IEE Archives and printed in F.H. Webb (Ed) *Extracts from the letters of the late Sir William Fothergill Cooke*, 1895.
2 Wemyss Reid, *Memoirs and Correspondence of Lyon Playfair, Lord Playfair of St Andrews*, 1899, p. 154.
3 Quoted in Geoffrey Hubbard, *Cooke and Wheatstone and the invention of the electric telegraph*, 1965, p. 135.
4 Archibald Geikie, *Life of Sir R.I. Murchison*, 1875.

Chapter 2

Early life

The earliest surviving record of Sir Charles Wheatstone is in the Baptismal Register of St Mary de Lode Church, Gloucester. It shows that Charles, elder son of William and Beata Wheatstone, of Trinity Parish was baptized on 17 March 1802.

There are two Wheatstones in the first Gloucester directory, which was printed in 1802 by Robert Raikes, the Gloucester printer and newspaper proprietor whose name is remembered as the pioneer of Sunday Schools. One was John Wheatstone, an attorney, and the other was his brother William, a cordwainer (shoemaker) and father of the future Sir Charles. William had a shop in Upper Westgate Street, now marked by an English Heritage blue plaque. Charles, the second child, was born on 6 February 1802 at the Manor House in Barnwood, which was then a village two miles from the centre of Gloucester along the Roman Ermine Street towards Cirencester. The Manor House was the home of his mother's parents, Samuel and Ann Bubb, members of an old Gloucestershire family. William's father John Wheatstone had also manufactured shoes, and other members of the Wheatstone family had been in business both in Gloucester and London as music publishers and musical instrument makers from about 1750. A third brother, Charles, had a musical business in London at 436 The Strand from about 1800 until his death.

In 1806 William and his family moved to London where he manufactured musical instruments, including flutes, and gave instruction on the flute and flageolet. He was proud to have among his pupils the popular Princess Charlotte (1796–1817), the only child of the Prince of Wales (later George IV) and Caroline of Brunswick. If Charlotte had not died in childbirth she would have become Queen on George IV's death in 1830. Instead her uncle became king, as William IV, to be succeeded in 1837 by her cousin Victoria.

Although only four when they moved to London, Charles had already been to a school near Gloucester and could read verses out of the Bible. In London he was sent to a school at Kennington kept by a Mrs Castlemaine, who was astonished at the progress he made under her care. He loved learning but was by

THE WHEATSTONE FAMILY

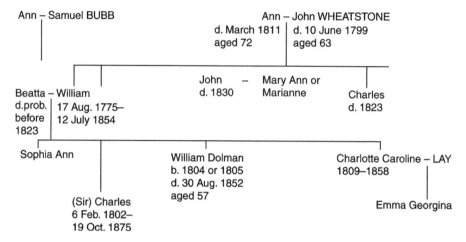

Figure 2.1   The Wheatstone family (see also Figure 14.2).

nature a nervous and timid child. Because of these traits he was inclined not to join in the sports of his schoolfellows and was regarded as unsociable.

At a later school he 'disputed with his teacher over what he was taught, which he considered inaccurate and deficient', and ran away. He got as far as Windsor but was brought back. From about 1813 William Wheatstone's business was at 128 Pall Mall, and Charles was sent to pursue his studies at a school in Vere Street, off Oxford Street. This school was run on the Lancastrian system in which the master taught the older boys, or monitors, who then taught the younger ones.

While at the Vere Street school one of his reports stated 'application moderate', yet he made considerable progress. In competition with older boys he won a gold medal for proficiency in French. He was then required to observe the school rule that the prizewinner should recite a speech at the prize giving, but this he absolutely refused to do, despite threats and coaxings, and he therefore forfeited the medal. As well as French, he learnt Latin and Greek, and he excelled in Mathematics and Physics.

In 1816 he was placed with his uncle Charles Wheatstone at 436 The Strand, but the uncle complained that young Charles neglected his work to pore over books, and preferred to shut himself up in an attic. Seeing the boy's interests, his father took him away from his uncle after a few months and encouraged him to pursue his studies with the aid of books borrowed from the library of the Society of Arts in the Adelphi, which William Wheatstone had joined in May 1815.

In 1817 when Wheatstone was 15, his activities included translating French poetry, writing some lines on the lyre which were later quoted in an engraving by the Italian engraver Gaetano Stephano Bartolozzi (1757–1821), and writing two songs. Thinking that his uncle would not look at his work, Wheatstone gave the songs to an established musician, Omera, who offered them as his own composition and the uncle appreciated and published them. Wheatstone often visited an old book-stall near Pall Mall and spent most of his pocket money on books which took his fancy – fairy tales, history and science. His introduction to electricity was an account in French of Volta's experiments. He bought it and, with the help of his brother, repeated many of Volta's experiments in the scullery of their father's house, using a home-made battery. Wheatstone also seems to have spent some part of his youth with his grandparents at Barnwood. His love of intricate mechanism and his ability as a showman were revealed when he set up automaton figures playing miniature musical instruments in the window of a shop in Gloucester, where 'Charley Wheatstone's clever tricks' were the talk of the town long afterwards.[1]

Wheatstone took a great interest in the science of sound, and made many experiments on the transmission of sound and the vibration of sounding bodies. He studied the comprehensive book on sound by the German physicist Chladni (1756–1827), and probably sought to emulate him. Years later, in a lecture given during his first session as a Professor at King's College London, Wheatstone quoted Chladni's statement of how he had come to take up the study of sound:

As an admirer of music . . . I remarked that the theory of sound was more neglected than most of the other branches of natural philosophy, which gave rise in me to the desire of supplying this defect.[2]

Chladni had demonstrated that a sounding body vibrated in a systematic manner. He scattered a fine layer of sand on a glass plate which was clamped at one point and made to sound by drawing a violin bow across its edge. The sand was found to leave some areas of the plate and gather in others, so that a symmetrical pattern was formed. Chladni argued correctly that the sand was shaken away from those parts of the plate where the vibration was greatest and accumulated where it was least. The pattern which appeared in the sand therefore revealed how the plate vibrated. Theorists, including Chladni, thought that the pattern of vibrations was in reality more complex than the pattern revealed by Chladni's sand figures: the areas in the sand figures ought to be subdivided by vibrations of smaller magnitude. These smaller vibrations could account for the differences in quality which can be observed in notes of the same pitch but from different instruments.

Wheatstone set out to make these smaller vibrations visible. He realized that the sand particles used by Chladni were not fine enough. He found that he could show the smaller vibrations in a horizontal plate by using a layer of water, instead of sand. Wheatstone achieved this independently, although he later found that it had been done by Örsted in 1813. As well as water, Örsted had used alcohol and lycopodion powder. Later, this work was to give the two men a common interest.[3]

His interest in the working of musical instruments was practical as well as theoretical, and he devised several new ones. The first was the keyed *flute harmonique*, which he made in 1818 though nothing is known about it. His first experiments on the transmission of sound arose from a consideration of the distance between the strings and sound boards of various instruments. In the piano it is about one centimetre, but in the violoncello it can be six or seven. To find how great a spacing could exist without adversely affecting the sound, he stretched a string on a steel bow and connected the bow to the sound board of the piano through a glass rod nearly two metres long. The sound was heard just as well as when the string was in direct contact with the board. The same result was obtained with a tuning fork. At a later date Wheatstone repeated this experiment in the theatre of the Royal Institution using a series of connected deal rods 12 metres long.

The Enchanted Lyre

Wheatstone's experiments on the transmission of sound were developed into a number of public demonstrations. The first, in his father's shop in Pall Mall, was the Acoucryptophone or Enchanted Lyre, which was first reported by several London journals in September 1821. The frontispiece to this book is an artist's impression of the scene.

Wheatstone enjoyed making up new words. Acoucryptophone is derived from the Greek and means literally 'hearing a hidden sound'. The visitor could see a device in the form of a large ancient lyre suspended from the ceiling and surrounded, but not touched, by a velvet covered hoop supported on the floor by three rods (or possibly hollow tubes, in the view of one mystified eye-witness). The horns of the lyre were like bugles bent down towards the floor and the discs on both sides of the body of the instrument were of metallic appearance. The lyre was suspended by a brass wire which passed through the ceiling and connected with the sound boards of instruments in a room above, where unseen players performed on the harp, piano and dulcimer. Only stringed instruments were used at first because Wheatstone found that, while it was comparatively easy to transmit in this way the sound of instruments having vibrating wires and sound boards, it was almost impossible to couple the lyre to instruments such as the flute in which the only vibration was that of a column of air.[4]

Regular concerts were given. The music critics urged music lovers to attend to hear the quality of the music produced and the skill of the unseen player or players. The strength of the lower notes particularly impressed them. Wheatstone overcame his shyness sufficiently to present the concert himself. He began by 'winding up' the lyre through a 'key-hole', but he never pretended that that served any useful function. He told his audience that the operation of the lyre was the application of a general principle for conducting sound. A future exhibition he was planning would conduct the sound of a whole orchestra.

This later exhibition opened in March or April 1822 in one of the shops along the Royal Opera Arcade. The admission charge was five shillings per head for an hour's orchestral concert, and here Wheatstone demonstrated both music and

Figure 2.2   *Wheatstone's drawing showing the Lyre hanging on a brass wire from the sound board on an instrument in the room above.*

voice conduction with his Diaphonicon, a horizontal sound conductor running between rooms. A further attraction was the 'Apparatus of the Invisible Girl', though unfortunately the anonymous author who wrote the report of Wheatstone's exhibition for Ackerman's Repository in 1822 did not feel it necessary to describe this. The Enchanted Lyre remained on show in various places at least until the summer of 1823.[5]

The family business

That autumn, when the future Sir Charles was 21, his uncle Charles died. Young Charles and his brother William Dolman Wheatstone took over their uncle's business in the Strand.

In 1829 there was extensive redevelopment of the eastern end of the Strand and the shop was demolished. The brothers found new premises at 20 Conduit

Street which became their home until Charles's marriage in 1847 and remained the address of the firm for the rest of his life. Their father had remained at 128 Pall Mall until 1822, then moved briefly to 24 Charles Street, St James's. He moved again in 1823 or 1824 to 118 Jermyn Street but about 1826 the father's business was amalgamated with his sons'. Charles and William continued to run the business and Charles continued to describe himself as a musical instrument maker even after his appointment as a professor at King's College London.[6] According to the contemporary *Post Office Directories* of London the occupants of 20 Conduit Street were known as 'Wheatstone & Co., Music sellers and publishers' until 1837 when they became 'Wheatstone, Charles & William, Musical instrument makers & music sellers'. From 1842 to 1847 the directories continued to name 'Wheatstone, Charles & William', thereafter it was 'Wheatstone, William & Co' until William's death in 1862, when it became 'Wheatstone & Co'. The trade description varied after 1842, but a typical one (1855) was 'Inventors and patentees of the concertina & manufacturers of harmoniums, music sellers & concertina makers'.

There seems little more to be known about Wheatstone's family. His father and grandfather had been skilled craftsmen. We do not know where they first lived in London, but since young Charles went to a school in Kennington it seems unlikely to have been the fashionable St James's where they lived later. Uncle Charles described himself as a music seller. In his will he left all his property, except for a few specific bequests, to be used to 'educate and support my nephews and nieces Charles Wheatstone, William Wheatstone, Ann Wheatstone and Charlotte Wheatstone, the sons and daughters of my brother William Wheatstone'. The nephews were to receive one quarter of the capital each on reaching 25 and the nieces on reaching 27. Charles was also to inherit 'my gold watch and its appendages'. Ann received 'my silver teapot and stand . . . silver sugar stand and tongs and half my silver spoons'. Charlotte had 'the remainder of my plate'. His brothers received £30 each and his sister-in-law Marianne 'my portrait hung over my drawing-room chimney place'. The only bequest outside the family was to 'my friend Miss Rosalind Martin of Kennington, as a token of esteem twenty-one pounds and three guineas for a mourning ring'.

Uncle John, the attorney, was married but childless. When he came to write his will he set out his wishes in four lines of verse:

> As to all my Worldly Goods now, or to be in store,
> I give to my beloved Wife, and her's for evermore.
> I give all freely – I no limit fix
> This is my Will and she's Executrix.

Having thus expressed his individuality, John then repeated his bequest in the conventional legal phraseology, and remarked that his legal experience had impressed upon him the importance of setting such things out clearly.[7]

Financially the family must have been comfortably off, although it is said the boys had to save their pennies to buy books. Certainly William Wheatstone could spare both time and money to devise improved instruments and he obtained a patent[8] in 1824 for 'a new method for improving and augmenting the tones of piano-fortes, organs and euphonons'. This invention consisted in covering as much as possible of the external surface of the instrument with a strong wooden frame and covering the frame on both sides with a tightly stretched vellum, parchment, paper, canvas or the like to form resonating chambers. The frame was arranged as closely as possible to the strings or pipes of the instrument, and could have holes in the outer covering with trumpet-like mouths.

Although William's somewhat impractical sounding invention bears little relation to Charles's work on sound conduction and free-reed instruments, it suggests that he shared his son's interest in the working of their instruments. Charles was not very interested in the commercial side of the business. Although he maintained some commercial interests throughout his life – and did quite well out of them – scientific matters were always dearer to him.

Notes

1 The main sources for Wheatstone's early life are: Roland Austin, 'Sir Charles Wheatstone, FRS', *The Gloucester Journal*, 10 October 1925; Anon, 'Memoirs of Deceased Members', *Minutes of Proceedings of the Institution of Civil Engineers*, 1877, pp. 283–90; Anon, 'Sir Charles Wheatstone, DCL, FRS', *Gloucestershire Biographical Notes*, c. 1894, pp. 188–200; G. Grove, *Dictionary of Music and Musicians*, 1879–89 and later editions; Anon, 'Scientific Worthies – VII. Sir Charles Wheatstone', *Nature*, 1876, pp. 501–3; Anon, 'Memoirs of Deceased Members', Proceedings of the Royal Society, 1876, pp. xvi–xxvii; Latimer Clark, President of the Society of Telegraph Engineers, remarks in opening the first meeting after the announcement of Wheatstone's death, *Journal of the Society of Telegraph Engineers*, 1875, pp. 319–34.

2 Quoted from the text in King's College library of Wheatstone's Sound Lecture V, given at the College on 17 March 1835. F-J Feltis, in his *Biographie Universelle des Musiciens et Bibliographie Générale de la Musique*, Paris, 1873, has a similar quotation (in French) in his article on Chladni.

3 M.C. Harding in the introduction to the Wheatstone section of his *Correspondence de H.C. Örsted avec divers savants*, 2 vols, Copenhagen, 1920.

4 Report 'Musical Intelligence – The Enchanted Lyre', *Ackerman's Repository*, September 1821.

5 Reports in *The Literary Gazette*, 15 September 1821 and *Ackerman's Repository*, March 1822.

6 See, for example, Wheatstone's English patent specification No. 7154 of 1836.

7 Wills in the Public Record Office: John Wheatstone, ref. PROB 11/1770; Charles Wheatstone, ref. PROB 11/1678.

8 English Patent No. 4994 of 1824.

PART 1
SOUND AND LIGHT

Chapter 3

Researches in sound

The Danish scientist Hans Christian Örsted (1777–1851) discovered that an electric current in a wire could deflect a compass needle. The discovery was announced in a short paper written in Latin in 1820, and it brought fame to Örsted because it showed for the first time that there was a connection between electricity and magnetism. During 1822 and 1823 Örsted made a tour of Germany, France and Britain, mainly to study the latest discoveries in optics. He arrived in London in May 1823, became acquainted with Wheatstone's work and saw the Enchanted Lyre.[1]

It is not known how Örsted and Wheatstone first met. Although Wheatstone was not then personally acquainted with the leading scientific men, his exhibition had attracted the attention of the press; Örsted, 25 years older than Wheatstone and an established scientist, may have been to see it as one of the sights of London. Örsted and Wheatstone found, in fact, that they had performed several similar experiments. When in Paris earlier that year Örsted had demonstrated experiments on Chladni's figures, and in June 1823 he wrote a letter to Wheatstone outlining his own early experiments on sound. This brief outline was included in Wheatstone's first formal scientific paper which Örsted encouraged him to write.[2] (This paper will be considered in the next chapter.) Örsted took it with him when he returned to Paris, and as a result Wheatstone's experiments were reported to the Académie Royale des Sciences on 30 June 1823. While in London Örsted had shown Wheatstone's experiments to several leading scientists and advised him to consult them.

On his return to Copenhagen, Örsted did not forget Wheatstone, but looked out for news of him and further researches by him. But Wheatstone published nothing for some years, and Örsted expressed his disappointment in a letter to the astronomer Sir John Herschel in July 1825. After receiving Örsted's letter Herschel got in touch with Wheatstone.

Subsequently, when Wheatstone was staying in Windsor, he wrote to Herschel at his home nearby in Slough:

The interest which you had taken in some former experiments of mine induced me to imagine that the continuation of them might have met with the same favourable reception from you, particularly as I have been fortunate enough to arrive at some, I venture to say, important inductions. One object of my calling upon you was to exhibit a little philosophical instrument which I think I shall call the Kaleidophone; it is the first I have constructed, and it displays with the greatest brilliancy imaginable the paths of vibrating rods; some experiments of this kind I have already shown you in a very imperfect and less varied manner. But my principal intention was to have submitted to your notice the results of several distinct series of multiplied experiments, which I am convinced will tend greatly to elucidate the real theory of undulating motion, and to establish more intimate relations between sound and light. I am extremely sorry that your indisposition has prevented me from laying these subjects before you, and from availing myself of your opinion thereon . . . I intended to see whether another series would interest you sufficiently to induce you to make a few confirmatory observations, which your extended apparatus would permit you more accurately to make, and your superior mathematical knowledge more aptly to apply.[3]

The letter bears a note in Herschel's hand to the effect that he had replied that he was not well enough to see Wheatstone then, but would be glad to see him and his experiments in London. Wheatstone was becoming known to the scientific establishment.

Wheatstone did call his new instrument the Kaleidophone, and he mentioned it when he wrote to Örsted in June 1826, explaining that the death of a near relative had forced him to devote his time to business affairs, though he expected soon to be free to resume his 'favourite pursuits'. (The near relative was his uncle Charles whose death a few months after Örsted's visit meant his nephews had to take over the business.) He said that he had been continuing his earlier experiments, and even extending his scientific interests. A paper on Vision was in preparation, and he would send Örsted a Kaleidophone if he would advise him how to dispatch one. An account of the Kaleidophone was published the next year, but his first paper on Vision was not published until 1838.

Wheatstone hoped to renew the acquaintance in a visit Örsted planned, but never made, in the spring of 1834. They met only once more, when Örsted came to the Congress of Naturalists in Southampton in 1846, but they corresponded and exchanged copies of their various papers.

However that first meeting came about, it was the paper Örsted took with him to Paris in 1823 which launched Wheatstone's scientific career. The paper was not read or published in full in Paris, but F.J.D. Arago gave an account of Wheatstone's experiments at a meeting of the Académie Royale des Sciences on 30 June 1823. The report of the meeting in the *Annales de Chimie et de Physique*, of which Arago was joint editor, includes all Wheatstone's experiments but omits most of his discussion. Possibly Arago did not agree with Wheatstone's ideas, while feeling that the experiments themselves were worth reporting. There is no indication of who Wheatstone was; the work was presented with the authority of Professor Örsted.

First scientific publication

Two months later the whole paper was published in England. It is the first article in the August 1823 issue of the *Annals of Philosophy* (also known as *Thomson's Annals*) and has the title 'New Experiments in Sound'. There is no editorial comment on the paper or its author, and no reference back to it by correspondents in later issues – though that is true of most papers in the *Annals*.

The paper is divided into three sections, 'On the Phonic Molecular Vibrations', 'Rectilinear Transmission of Sound', and 'Polarization of Sound'.

The first section begins with an analysis of the various kinds of 'phonics' – the term Wheatstone uses for 'those bodies which, being properly excited, make those sensible oscillations which have been thought to be the proximate causes of all the phenomena of sound'. He recognizes three main groups of phonics, which he called 'linear phonics', 'superficial phonics' and 'solid phonics'. Most of the phonics he considers come in the first group, and these divide into 'transversal linear phonics' and 'longitudinal linear phonics'. The transversal group include stretched strings (as in a piano or violin) and rigid devices such as tuning forks; these all make oscillations at right angles to their axis. The longitudinal group make oscillations in the direction of their axis: examples of this group are columns of air (as in a flute) and solid rods. Superficial phonics include stretched surfaces (as in a drum) and rigid surfaces (for example bells and sounding plates). The group 'solid phonics' includes volumes of gas, but Wheatstone gives no specific example and may have included the category merely to be comprehensive. Any of these phonics can be made to oscillate rapidly and produce sound, said Wheatstone, and the phonic may oscillate either as a whole or with several subdivisions separated by stationary points or lines called nodes, or nodal lines. Chladni had made visible the oscillations of a particular phonic, a sounding plate, by strewing sand on it. Wheatstone described this and also referred to Chladni's thought that more minute vibrations must be present in order to account for the varieties of quality or 'timbre' which can be found in notes of the same pitch. Wheatstone then explained how he had demonstrated the existence of these more minute vibrations by covering a sounding plate with a thin layer of water in place of the sand used by Chladni. When the plate was made to sound by drawing a violin bow across an edge, the water developed what Wheatstone described as 'a beautiful reticulated surface of vibrating particles'. This was proof that the motion of a sounding plate was more complex than Chladni had succeeded in showing, and it was this experiment which Arago described first in his account of Wheatstone's work to the Académie.[4]

Wheatstone used the expression 'phonic molecular vibrations' for his newly demonstrated particle movement. In 1823 the word 'molecule' could legitimately be used to mean any small particle, and the adjective 'molecular' came into the language in that year. Wheatstone was not the first user of 'molecular', but he was the first person to use 'phonic' as an adjective and as a singular noun. The plural noun 'phonics', meaning the science of sound, was in use from the seventeenth century and still in current usage when Wheatstone was writing, although it has since become obsolete. Wheatstone also demonstrated 'phonic molecular vibrations' in a sounding column of air. If the open end of a flute or flageolet is placed on the surface of a vessel of liquid – or, better still, slightly raised from the surface but connected by a thin film of liquid – the vibrations are visible on the surface of the liquid.

At this point in the paper there is an interesting note:

I, however, conceived I was the first who had indicated these phenomena by experiment, until a few days ago, repeating them, together with others which form the subject of this paper, in the presence of Professor Örsted of Copenhagen, he acquainted me with some similar experiments of his own. Substituting a very fine powder, Lycopodion, instead of the sand used by Chladni for showing the oscillations of elastic plates, this eminent philosopher found the particles not only repulsed to the nodal lines, but at the same time accumulated in small parcels on or near the centres of vibration; these appearances he presumed to indicate more minute vibrations, which were the causes of the quality of the sound. Subsequently he confirmed his opinion by observing the crispations of water or alcohol on similar plates, and showed that the same minute vibrations must take place in the transmitting medium, as they were equally produced in a surface of water when the sounding plate was dipped into a mass of this fluid. These experiments were inserted in Lieber's 'History of Natural Philosophy', 1813.

Wheatstone did not explain how he, a young and virtually unknown instrument maker, came to be conducting experiments in the presence of the internationally famous professor.

The second section of the paper is concerned with the transmission of sound through linear conductors. Wheatstone refers to his Enchanted Lyre, and states that linear sound conductors possess nodes similar to those in linear phonics. The vibrations of high pitched sounds can be transmitted through thinner stretched wires than the vibrations of low sounds. In the third and final section of the paper Wheatstone states that sound transmitted in a linear conductor may be 'polarized', and he demonstrates that a right-angle bend in a linear conductor will either pass or stop the sound depending on whether that conductor is bent in the plane of the sound vibrations or in a plane at right-angles to the vibrations. If a harp string is made to vibrate in a direction at right-angles to a sound board and the vibrations are transmitted to the sound board then the note is heard. But if the string vibrates parallel to the board there is little or no transmission and the sound is barely heard. Wheatstone discussed the analogy which he believed existed between sound and light. If the polarization of light was a similar phenomenon to his polarization of sound there was no need to assume that 'reflecting surfaces act on the luminous vibrations by any actual or repulsing force, causing them to change their axes of vibrations; the directions of the vibrations in different planes, as I have proved exist in the communication of sound, are sufficient to explain every phenomenon relative to the polarization of light'.

Arago expressed reservations about the analogy Wheatstone drew between the phenomena of sound he described and the polarization of light. In the French report Arago added a footnote to the section 'polarization of sound', saying that he had kept the original title but he did not share Wheatstone's ideas.

Wheatstone's concept of polarization of sound caught the interest of other scientists because of its relevance to the theory of light. There were two conflicting theories of the nature of light: the corpuscular theory of Newton, and Huygens' wave theory. Newton's view that light was a substance, with particles subject to the laws of mechanics, was widely accepted until the early years of the nineteenth century. However, after 1800 Thomas Young took up the wave theory and developed the idea that light was a vibration in an all-pervading ether. He thought the vibrations were longitudinal, but the discovery of polarization by E-L. Malus in 1808 forced him to accept that the vibrations were transverse and that in polarized light the vibrations were restricted to one plane. Young's ideas were strongly opposed. Malus and other supporters of the corpuscular theory explained polarization by saying that particles of light were not perfectly spherical but possessed some asymmetry which could be described as the presence of poles – hence the term 'polarization'. The eventual victory of the wave theory was largely due to the work of the French engineer A-J. Fresnel who, between 1815 and 1824, explained mathematically in terms of the wave theory the reason why light travels practically in a straight line, and the phenomena of diffraction and polarization. Even Fresnel's ideas were not always accepted and Arago was strongly opposed to them. Any experiment which might help to resolve the problem of the nature of light was therefore of great interest at the time. Supporters of the wave theory would immediately see the analogy Wheatstone was drawing between sound and light. But it was only an analogy – albeit a good one – and supporters of the corpuscular theory of light could decline to accept it, as Arago did, and say that light and sound were quite different.

Arago's account of Wheatstone's work was translated into German by L.F. Kaemtz and published the following year in *Schweigger's Journal*. Kaemtz retained Arago's note of reservation, and added a long 'postscript' of his own in which he said, 'The most interesting thing about this dissertation is Wheatstone's apparent discovery of the Polarization of Sound'. He then quoted some experiments by Herr Muller, of Breslau, who claimed to have observed polarization phenomena when the handle of a tuning fork was coupled to a sounding board through a crystal of calcspar. Kaemtz had tried without success to repeat Wheatstone's and Muller's observations. He appears to have been very sceptical about the whole matter and twice commented that such variations in intensity of sound as he did hear might well have been caused by his holding the tuning fork slightly differently.

Two years later, in 1826, *Schweigger's Journal* published another article on the polarization of sound. This was by Wilhelm Weber and was translated into English and published under the title 'On the Polarization of Sound in a different manner from that described by Mr Wheatstone' in the *Edinburgh Journal of Science* for June 1826. Wheatstone had a copy of the *Edinburgh Journal* for that month and the page reference is marked boldly on the cover in his own hand. The article refers throughout to a 'pitchpipe', though it is clear from the description that a tuning fork is envisaged, and Wheatstone has crossed out the word 'pitchpipe' and written in 'tuning fork' in his copy. Later when Wheatstone referred to the tuning fork in one of his papers he made a point of describing it. Weber's article makes the point that if a tuning fork is struck and held near the ear the intensity of sound heard varies if the fork is rotated on its axis of symmetry. If one of the broad sides or one of the narrow sides is directed to the ear the intensity of sound heard will have a peak value, and in between these four positions there are four positions of minimum sound. Weber reports an experiment of Chladni's in which a glass vessel was tuned (by adding water) to the same note as a tuning fork. When the fork was struck, held to the mouth of the vessel and rotated, the intensity of sound was heard to pass through four maxima and four minima in one complete rotation of the fork. Kaemtz had noticed an apparent variation in the sound from his tuning fork as he turned it, and had attributed it to the way his fingers gripped the fork. Weber (or possibly Chladni, since Weber was partly quoting Chladni's work) had shown that the intensity of sound was in fact determined by the angular position of the fork.

It is reasonable to suppose that Wheatstone studied Weber's paper and repeated the experiments for himself. The idea of polarization of sound is not mentioned again in Wheatstone's published work, though some of the experiments which led him to the idea are repeated in a paper of 1831. It may be significant that, in the two years following the publication of Weber's paper, Wheatstone investigated the phenomena arising when a tuning fork is applied in various ways around the head and the manner in which a column of air resonates in response to a tuning fork. This work was made public through the Royal Institution.

The Royal Institution

About 1825 Wheatstone began an association with the Royal Institution which lasted for the rest of his life, and he became a close friend of Michael Faraday. It was in that year that Faraday was made 'Director of the Laboratory' at the Institution, and started the regular weekly evening meeting of members which still continues as the Friday Evening Discourse. There is nothing to indicate whether Wheatstone was introduced first to Faraday and then to the Royal Institution, or first to the Institution, but he did not himself become a member of the Institution until 1846. It may be that Örsted's introductions were the starting point of this association, but, as we have seen, Wheatstone was becoming known for his researches on sound.

However Wheatstone and Faraday first met, the association was to prove invaluable for Wheatstone. He was incapable of giving a good public lecture himself, but Faraday, who loved music and was intrigued by Wheatstone's researches into the nature of musical sound, put his lecturing talents at Wheatstone's disposal and gave a number of discourses for him.

There is a much repeated story about Wheatstone and the Royal Institution which may well be mentioned here. It is said that Wheatstone had agreed to give a Friday Evening Discourse at the Royal Institution himself, but at the last minute his usual shyness overcame him and he ran down the stairs and fled from the building. In his biography of Faraday, Professor L. Pearce Williams relates the story thus:

It was on Friday, 10 April 1846, that Sir Charles Wheatstone bolted down the stairs of the Royal Institution, leaving Faraday to deliver the discourse. Almost unprepared for such an eventuality, Faraday was able to fill up only part of the evening with the announced subject and so, to complete the talk, he revealed his thoughts on the nature of light. These speculations, entitled 'Thoughts on Ray-vibrations' completed his general concept of the unity of force and of the universe.

The story is probably founded on fact, but has grown with much re-telling. The Royal Institution's Friday Evening Discourse programme covering April 1846 reveals that no discourse was planned for 10 April (it was Good Friday) and also that no subjects or speakers but just dates were announced in advance for the session. Among Faraday's lecture notes in the Royal Institution Archives is a sheet of paper with the heading 'Wheatstone's Chronoscope' and the date 3 April 1846. Included at the end was a section with his thoughts on 'Ray vibrations'. The speaker originally planned for the evening was James Napier, who had to withdraw for reasons that are not clear, but it was not a last-minute cancellation. *The Athenaeum*'s list of 'Meetings for the Ensuing Week', dated 28 March, shows Faraday as the speaker on 3 April.

Nevertheless, the story is now part of Royal Institution legend, often alluded to by speakers, who say they have been shut in with the Director before their Discourse to prevent them from 'doing a Wheatstone'. The present writer's experience is that the lecturer is not 'shut in', but is very hospitably entertained.

The first recorded occasion on which Faraday spoke for Wheatstone was on 15 February 1828. The subject of the lecture was 'Resonance', and the account in the Proceedings of the Royal Institution[5] states that it was 'delivered by Mr. Faraday, who, however, gave all the credit belonging to the illustration, and the new information communicated, to Mr. C. Wheatstone'. The lecture was illustrated by experiments on Javanese musical instruments. These were lent by Lady Raffles from the collection brought back by Sir Stamford Raffles (1781–1826) who, while in various government posts in the Far East between 1805 and 1824, made extensive studies of the countries in which he served. This lecture, together with some supplementary material delivered on 7 March, was the substance of Wheatstone's paper 'On the Resonances or Reciprocated Vibrations of Columns of Air', which is considered below. There is no record of the audience's reaction to the first discourse, but it was presumably favourable since Faraday gave several more with material supplied by Wheatstone. In Faraday's manuscript index of discourses in the Royal Institution archives these are all listed as being by 'Faraday and Wheatstone', whereas all Faraday's other discourses are stated simply to be by 'Faraday'. He never gave a discourse on behalf of anyone else.

At the Institution on 9 May 1828 the subject was the 'Nature of Musical Sound'. Again, the report in the Proceedings states that 'it was delivered by Mr. Faraday, but supplied by Mr. Wheatstone'. There was no written paper published after this lecture, but the report in the Proceedings summarizes it in about one thousand words. The lecture considers the nature of pitch, the upper and lower frequency limits of audibility, the production of sound by friction and by a card striking teeth on a revolving wheel, and the production of sound by a siren. Finally it considers the action of the recently invented 'Mund Armonica' – the mouth organ, from which Wheatstone was soon to develop the symphonium and the concertina. Except for this final section the lecture appears to be very similar to the first lecture on sound which Wheatstone gave when he became Professor at King's College.

The Kaleidophone

Before the date of the first lecture which Faraday delivered on his behalf, Wheatstone's name had appeared in the 'Proceedings of the Royal Institution' for 1827 as the author of a paper, 'Description of the Kaleidophone or Phonic Kaleidoscope: a new Philosophical Toy, for the Illustration of Several Interesting and Amusing Acoustical and Optical Phenomena'. There is nothing to suggest that the substance of this paper was ever given as a lecture. The Kaleidophone was the instrument Wheatstone had mentioned in a letter to Örsted in 1826 and also sought to demonstrate to Herschel the previous year.

In the paper Wheatstone comments that the application of scientific principles to ornamental and amusing purposes increases their popularity and helps one to remember the effects. He explained his choice of the name 'Kaleidophone' by saying that the instrument resembled Brewster's Kaleidoscope in that it 'creates beautiful forms', but there was no other similarity. The Kaleidoscope was devised by Sir David Brewster in 1817. The Kaleidophone was inspired by the work of Thomas Young who wound a fine silvered wire in a spiral on a low string of a square piano and observed it with light from a slit when the note was sounded. The result reminded Young of Chladni's figures but he thought his results might be within the reach of mathematics. Six years later Wheatstone was to produce an explanation of Chladni's figures in a mathematical form. Possibly he hoped to interpret mathematically the movements of a sounding body with the aid of the Kaleidophone, but if that was his intention he did not succeed. His Kaleidophone never provided more than a visual demonstration of the complex motion of a sounding body.

In its simplest form the Kaleidophone is a metal rod clamped at one end and having a small reflector fixed on the free end. If the rod is struck with a padded hammer or excited with a violin bow, and the bead observed by means of a point source of light, such as a candle, then regular patterns can be seen as the rod vibrates, carrying the bead with it. Because of the persistence of vision the observer sees a pattern of lines, not a moving point of light. The instrument

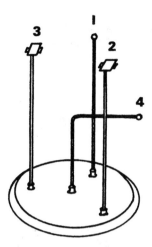

Figure 3.1 The Kaleidophone (from Wheatstone's drawing).

described in the paper was more elaborate, with four rods each about 30 centimetres long fixed in a wooden base. One rod was cylindrical, two or three millimetres in diameter, and with a small silvered glass bead cemented onto the free
end. This rod gave the patterns shown in Figure 3.2. Another rod had a hinged
plate on its end, instead of the bead. This might be used for carrying several
beads, in which case more elaborate and attractive patterns would be formed. If
the plate was vertical and arranged to carry a flat object, the object might appear
solid as it vibrates. The third rod had a square cross-section. If made to move in
the direction of one of its sides it would oscillate in a straight line, but if made to
move obliquely the motion obtained was the resultant of two separate motions
at right angles. A point on a straight rod could only move in a plane. Wheatstone
therefore added a fourth rod, bent through a right angle. When suitably struck
the two parts oscillated separately, and the reflecting end described a three-
dimensional figure. Wheatstone noted in passing that any regular curvilinear or
angular motion of not too great amplitude might be shown in a similar way. If
an object such as a word printed on a card was placed on one of the rods and
vibrated, two distinct images were seen connected by a blur.

Wheatstone added an experiment to demonstrate the persistence of vision
upon which the Kaleidophone depended. An opaque disc having a sector cut
out was rotated in front of a picture and the whole picture could be seen. The
only accurate experiments to measure the duration of the persistence of vision
before that time were made by d'Arcy, who published a memoir on the subject in
1765. Wheatstone did not use the phrase 'persistence of vision', but 'the apparent duration in the same places of visible images after the objects which excite
them have changed their positions'. Three Kaleidophones were on display at the
Friday evening meeting of the Royal Institution on 4 May 1827, when the main
business of the evening was a lecture by Faraday on chlorine and its compounds,
followed by the opening of a Mummy of a Cat by Dr Granville.

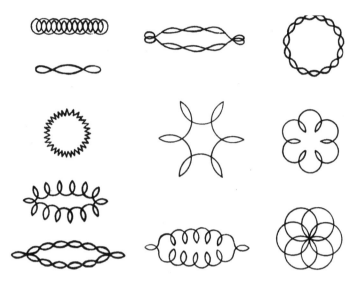

Figure 3.2   Patterns produced by the Kaleidophone (from Wheatstone's paper).

Resonance

Later the same year Wheatstone published a paper entitled 'Experiments on Audition'. In the nineteenth century 'audition' was synonymous with 'hearing', and this paper was concerned with the mechanism by which sounds were heard. Wheatstone referred to recent experiments by Savart and Wollaston and observed that although the physiology of the ear had been studied for centuries the functioning of the various parts was not fully understood.

The effect of covering the ear with the hand or closing the outer ear passage by the finger without pressure is that the sounds of one's own voice or a tuning fork applied to the head are augmented and are heard mainly by the closed ear. The reedy sound of one's own voice in this experiment is removed by compressing the air in the meatus against the eardrum or by exhausting the Eustachian tube. In either case the eardrum is put into slight tension. The Eustachian tube, which runs between the pharynx and the inner ear, can be exhausted slightly by closing both mouth and nostrils and then sucking. The augmentation is due to 'reciprocated vibrations' (that is, resonance) as may be proved by blocking the ear with wool, when the effect is not obtained. The same effect as in blocking the ear is obtained by filling the meatus with water.

In a footnote to the paper Wheatstone described the tuning fork, which he found a very convenient instrument for a variety of acoustical experiments:

The tuning fork consists of a four-sided metallic rod, bent so as to form two equal and parallel branches, having a stem connected with the lower curved part of the rod, and contained within the plane of the two branches . . . The sound produced by this instrument when insulated is very weak . . . . but instantly its stem is connected with any surface capable of vibrating, a great augmentation of sound ensues from the communicated vibrations.

The tuning fork had been used by musicians for over one hundred years, although according to the *Oxford English Dictionary* that name only dated from 1799. It is interesting that Wheatstone felt it necessary to describe its construction in such detail for his scientific readers.

Wheatstone then described an instrument which he called a 'microphone' (see Figure 3.3). This had two flat pieces of metal each large enough to cover the external ear. A rod of iron or brass wire 40 centimetres long and two millimetres in diameter was riveted to each plate and the rods were joined at their other ends so as to meet in a single point. The microphone could be held to the ears just by the springiness of the rods or by a ribbon tied round the head. Wheatstone suggested four experiments using the microphone. If a bell was rung in a vessel of water the difference of intensity at different distances from the bell was easily detected; if the point of the microphone is applied to the sides of a vessel containing a boiling liquid sounds were heard very distinctly; the points of greatest or least vibration in a violin or guitar were easily found; and lastly, if a sounding tuning fork was in contact with any part of the microphone and at the same time a musical sound was produced by the voice, then 'the most uninitiated ear will be able to perceive the consonance or dissonance of the two sounds'.

Wheatstone does not mention the doctor's stethoscope, which was first a roll of paper, then later a wooden rod. It was devised by R.T.H. Laënnec (1781–1826) in 1816 and described in his book, *De l'auscultation médiate*, published at Paris in 1819. The subtitle of Laënnec's book explained that it was about investigating disorders of the heart and lungs using a novel method of investigation. There is no copy in Wheatstone's library, and he may not have been aware of Laënnec's work .

The substance of the first discourses which Faraday gave for Wheatstone at the Royal Institution was quickly published under the title 'On the Resonances, or Reciprocated Vibrations of Columns of Air', in *The Quarterly Journal of Science* for the first quarter of 1828. In 1832 Wheatstone himself lectured at the Royal Institution 'On the Vibrations of Columns of Air in Cylindrical and Conical Tubes'. There was a report of this lecture in *The Literary Gazette* which

Figure 3.3   *Wheatstone's 'Microphone', used like a doctor's stethoscope to hear faint sounds.*

included the comment that 'On a subject of this kind, where so much depends on experiments, it must be manifest that no report of ours can convey an adequate idea of its interesting nature'.

When Professor William Grylls Adams addressed the Musical Association on the subject of Wheatstone's 'Musical Inventions and Discoveries', a few months after Wheatstone's death, he spoke mainly on the subject of resonance. His concluding remark was that 'the whole theory of harmonics when applied to wind instruments might be said to be made out by Sir Charles Wheatstone'.[6]

Wheatstone's paper 'On the Resonances . . .' begins with a brief explanation of the fact that any body which can vibrate may be set in motion by small but repeated pulses. He quotes Galileo's observation that a heavy pendulum may be put in motion by the least breath, provided the blasts are repeated and keep time with the vibrations of the pendulum. The column of air in a flute will resonate if a vibrating tuning fork is brought near and the flute is appropriately fingered. The experimental basis of the paper is evident from the remark at this point that if a C tuning fork is employed then the flute must be fingered for B, since in normal use the player's lips flatten the note of the flute by a semitone. The flute will also resonate to a sound produced by a wind instrument, such as another flute, played nearby.

A cylindrical or prismatic column of air in a tube open at both ends may vibrate not only as a whole but in any number of equal subdivisions. It will therefore resonate to sounds whose frequencies are the fundamental frequency multiplied by 1, 2, 3, 4, 5, etc. A tube closed at one end will resonate only to sounds whose frequencies are the fundamental frequency multiplied by 1, 3, 5, 7, etc. If two or more different sounds occur together then one sound may be made separately audible by a column of air resonant at the appropriate frequency. The 'third sound' or 'grave harmonic' heard when two harmonious sounds occur together cannot be emphasized by a suitably tuned resonant column of air because that column will necessarily be resonant also to at least one of the original two sounds. The third sound is of a frequency equal to the difference between the frequencies of the two original sounds. Wheatstone found this result experimentally, but then explained it mathematically.

He discussed the application of resonant columns of air in practical musical instruments. Among the Javanese musical instruments brought to England by Sir Stamford Raffles was the Gender (see Figure 3.4). This had 11 metal plates held on two horizontal strings, arranged as in a xylophone, and 11 bamboo tubes arranged vertically beneath the plates with their upper ends open. The tubes were of different lengths and each one resonated at the lowest natural frequency of the corresponding plate. Wheatstone placed a board between the plates and the tubes and showed that when struck each plate produced a fairly high-pitched note. When the board was removed, however, and a plate struck, then the column of air in the corresponding tube resonated and the instrument produced a rich deep tone.

The demonstration of the Gender was followed by an experiment with a tuning fork and a single tube whose effective length could be varied with a

Figure 3.4 *The Gender, an Asian musical instrument. The horizontal bars act as a xylophone and vertical bamboo tubes beneath the bars provide resonating chambers to amplify the sounds.*

piston. When the length of the air column in the tube was such that its natural frequency was the same as that of the fork, then it resonated when the fork was struck, amplifying the sound. If the length of the tube were varied, by moving the piston, there was initially no resonance, but if the length were halved, or divided by two, three, or any whole number, then the tube again resonated but at twice, three times, etc. the frequency of the fork.

Wheatstone commented that although several Asiatic and African musical instruments used the resonance of a column of air to boost the sound of another vibrating element, there was no European instrument that did so. He intended to publish another paper on the subject and describe ways he had devised in which the principle might be put to practical use. No such account was ever published, but Wheatstone did announce a new instrument (the Terpsiphone, described later) using the principle; nothing further was heard of it.

Another instrument which depends on the resonance of an air chamber is the Jew's harp, or guimbarde. Wheatstone felt it necessary to describe this for his Royal Institution audience. It has an elastic steel tongue riveted at one end to a brass or iron frame which can be held in the teeth. The free end of the tongue is bent at a right angle, making it easy for the player to strike with a finger. The resonant air chamber is the player's mouth. The fundamental frequency of the Jew's harp is lower than the lowest sound to which the mouth can resonate, so the sound that is heard is always a harmonic. This instrument illustrates the principle enunciated above that resonance may occur at a harmonic frequency, for that is the only way in which it can operate. The original discourse was illustrated with a demonstration of the Jew's harp by a Mr Eülenstein (see Figure 3.5). Wheatstone remarked that 'Those who have heard only the rude twanging to which the performance of this instrument in ordinary hands is confined, can have no idea of the melodious sounds which, under Mr Eülenstein's management, it is capable of producing'.

The subject of resonance was considered further in a discourse Faraday gave for Wheatstone on 3 April 1829. No published paper followed this discourse, which was illustrated with a demonstration by a Mr Mannin who had the power of producing two sounds simultaneously and could whistle a duet. Wheatstone did not reveal where he found such remarkable performers to illustrate his subject, and Messrs Eülenstein and Mannin can rarely have had so distinguished an audience or found themselves the centre of scientific interest!

The 1832 lecture 'On the Vibrations of Columns of Air . . .' described a continuation of the work in the paper just considered. The main new result was his experimental demonstration of the way in which the air in a tube vibrated when the tube was open at both ends and producing the lowest sound of which it is capable. According to Bernouilli's 'Theoretical Investigations', stated Wheatstone, the motions of the particles of air at both ends of such a tube were simultaneously towards and away from the central point or node. Wheatstone used for his demonstration a tube bent almost into a circle so that its ends were opposite each other with a small space between them. He took a glass plate capable of sounding the same note as the tube and placed it between the ends of a tube. On drawing a violin bow across the plate to make it sound, the vibrations at the plate were necessarily towards one end of the tube and away from the other. If the theory were correct the effects at the two ends of the tube would neutralize each other and the tube would not resonate.

Figure 3.5   The Jew's Harp.

Wheatstone demonstrated that that was the case and therefore the theory was correct.

He then showed experimentally that, when a column tube resonated at a higher note than its fundamental, the column was effectively divided into a number of equal parts and that the tube could be open to the atmosphere at the junctions of the parts without affecting the sound. He showed that a solid partition could be inserted at the centre of an open tube without affecting the fundamental sound, therefore the air at the centre must be at rest. He disproved the view, which he said had been stated by Chladni among others, that 'the end at which a tube is excited into vibration must always be considered as an open end, even if it be placed immediately to the mouth'. He demonstrated the phenomenon of sounding gas flames and showed that the loudest tone is produced in the smallest tubes, which was contrary to the general opinion. Finally he considered the harmonics of conical tubes, showing that they are similar to the harmonics of cylindrical tubes of the same length.

At the British Association meetings in 1835 Wheatstone repeated his experimental demonstration of the correctness of Bernouilli's theory. He used a lead tube about 2.5 centimetres in diameter and 30 centimetres long, bent into a circle, and on this occasion the tube had a joint in the middle which enabled the ends to be offset from each other. He showed that if the two ends of the tube were adjacent to portions of the vibrating plate that moved in opposite directions, both towards or both away from the ends of the tube, then no resonance occurred.

Wheatstone made a major contribution to acoustics by his continuing study of Chladni's figures. Although Professor Adams selected the work on the theory of harmonics and resonance for the main part of his memorial address on Wheatstone's musical inventions and discoveries, the writer of the obituary in the Proceedings of the Royal Society remarked that 'his principal contribution to Acoustics is a memoir on the so-called Chladni's figures ... probably the most remarkable of his early scientific labours'.

The substance of the memoir, which was published in the *Philosophical Transactions of the Royal Society* in 1833, was a lengthy and thorough analysis of some of the more complex Chladni's figures that could be produced on a rectangular sounding plate (see Figure 3.6). Wheatstone showed that all the figures, however complex, could be explained by superimposing a number of simpler patterns. He calculated the effect of the superposition by a simple process involving no mathematics, and he repeated the process for enough examples to leave little doubt that the rule he had demonstrated in a few specific cases was a universal one.

The paper on Acoustic Figures was Wheatstone's last publication on the theory of sound. By this time (1833) he was becoming interested in electricity, but before considering that subject we will look at Wheatstone the musical instrument manufacturer, who had been applying some of his scientific discoveries in the course of his work.

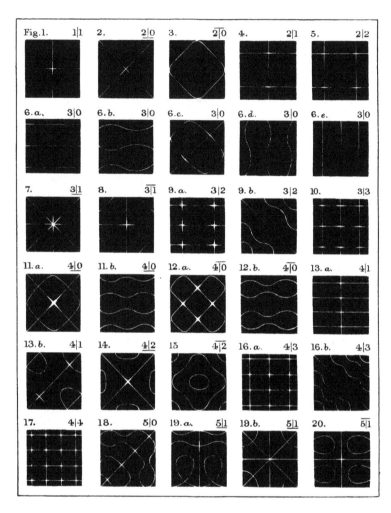

Figure 3.6   Chladni figures. When a horizontal plate is made to vibrate, producing a sound, some parts of the plate are in rapid motion while other parts are static. Chladni demonstrated the pattern of vibrations by scattering a layer of sand on the plate. When the plate vibrated, the sand was then shaken towards the points of least movement. Wheatstone used a finer powder than sand and was able to show finer details in the pattern.

Notes

1   M.C. Harding in the introduction to the Wheatstone section of his *Correspondence de H.C. Örsted avec divers savants*, 2 vols, Copenhagen, 1920.
2   Wheatstone's published scientific papers are listed in the Appendix.
3   Letter, Wheatstone to Herschel, 23 August 1825, in Herschel papers, Royal Society Archives.

4 'Extrait des Scéances de l'Académie Royale des Sciences', *Annales de Chimie,* **23**, Paris, 1823, pp. 310–13.
5 'Proceedings of the Royal Institution', in *Quarterly Journal of Science,* 1828, p. 173.
6 W.G. Adams, 'On the Musical Inventions and Discoveries of the Late Sir Charles Wheatstone, FRS', *Proceedings of the Musical Association,* **2**, 1876.

Chapter 4

Musical instrument manufacturer

It was characteristic of Wheatstone that he was always alert both to the scientific lessons which might be learnt from everyday things and to the practical applications of scientific discoveries. For that reason the division of subject matter between the previous chapter on Wheatstone's researches in sound and this chapter on Wheatstone as a musical instrument maker is somewhat arbitrary. He studied the transmission of sound because he was interested in the working of the instruments he made, and in particular the process by which sound created by the vibration of the strings of a piano or violin is transmitted to the sound board. His work on the transmission of sound, and also that on the development of an artificial voice, may be regarded either as pure research or as a potentially viable commercial venture which was superseded by electrical communication.

The transmission of sound

From the accounts of Wheatstone's youthful researches it is clear that the transmission of sound through solid conductors had already interested him for a long time when in 1831 he published 'On the Transmission of Musical Sounds through Solid Linear Conductors, and on their subsequent Reciprocation'. The substance of this paper had earlier been given as a Royal Institution Discourse by Faraday on 5 March 1830. The anonymous writer in *Ackerman's Repository* who had described Wheatstone's demonstration of sound transmission in 1821 had speculated upon the future development of Wheatstone's work. The report envisages music being transmitted across London from one hall to another, or laid on to homes, and parliamentary debates being heard at once 'instead of being read the next day only'. This speculation almost certainly originated with Wheatstone himself. When writing in 1840 about his early telegraph work[1] he claimed to have been interested in establishing communication between distant places from the time his early experiments were reported in *Ackerman's Repository.*

In the 1831 paper Wheatstone gave the results of many experiments and discussed the practical limitations to the transmission of sound over long distances. He observed that if a conducting substance could be obtained with perfectly uniform density and elasticity then it would be as easy to transmit sound from London to Aberdeen as from one room to the next. He said that from experiments by Peyrolle it appeared that the intensity with which sound was transmitted through a solid conductor was nearly proportional to the velocity of transmission. There was no mention of electricity in Wheatstone's paper, but it was already established that the velocity of electricity was far greater than that of sound in any substance. It may be significant that his electrical researches began with the measurement of the velocity of electricity and led from this to an electric telegraph. The paper ended with the suggestion that future progress lay in developing an artificial voice, which could produce speech and communicate it to a conductor with greater volume and greater efficiency of coupling than a human voice. He believed that such a 'voice' would soon be made.

Wheatstone only considered the transmission of speech or music: the idea of signalling by code using, for example, a series of loud bangs, does not seem to have occurred to him at that time. Later, with the electric telegraph, he was reluctant to use any form of code and always sought after a direct reading system.

Talking machines

The distance that sound could be transmitted was limited. An alternative way of transmitting speech to a distance might be to use a talking machine, controlled from a distance. The Hungarian, Wolfgang von Kempelen, built a talking machine in Vienna and described it in a book published in both French and German editions in 1791. Wheatstone built his own machine, developed from von Kempelen's, and described it at the British Association's meetings in 1835. At the end of his demonstration Wheatstone gave 'an analysis of the elements of speech founded on these and other investigations, and pointed out the importance of the inquiry as connected with philology'. The concept of a single Indo-European language from which most European languages were derived was fairly new, having been postulated by Franz Bopp in 1816. By 1836 such a well-read scholar and linguist as Wheatstone would have been aware of the comparative studies of grammar and vocabulary in different Indo-European languages. There is no evidence, however, that Wheatstone's artificial voice or anything similar played any part in the development of philology. In 1837 Wheatstone published a long review article on speaking machines, beginning with the 'pretended speaking machines' of ancient and medieval times in which statues were caused to appear to speak with the aid of speaking tubes.

Although Wheatstone did not develop his speaking machine any further, he kept it at the College for the rest of his life, and was prepared to demonstrate it on request. The actor W.C. Macready, who several times included Wheatstone among his dinner guests in the 1840s, recorded in his diary that on 29 April 1840 he saw Professor Wheatstone at King's College demonstrating the electric telegraph and a speaking machine which said, 'Momma, papa, mother, thumb, summer'. Twenty years later, Alexander Graham Bell, the inventor of the telephone, spent a year in London when he was 13 and on one occasion visited Wheatstone with his father, an elocutionist with a special interest in the mechanism of speech. Bell recalled later, 'I saw Sir Charles manipulate the machine and heard it speak, and although the articulation was disappointingly crude, it made a great impression upon my mind'. Perhaps having heard a machine speak encouraged Bell in his work towards the telephone.[2]

The premises at 20 Conduit Street, to which the brothers Charles and William Wheatstone had moved their home and business in 1829, remained in the family for the rest of Charles Wheatstone's life. His active involvement in the business seems to have ended about the time of his marriage in 1847, when he moved to Hammersmith, but resumed after his brother's death. In 1872 he was responsible for the rent of 20 Conduit Street, for there is a note among his papers at the College asking for a quarter's rent up to 25 March 1872, though the amount is not specified.

New musical instruments

Wheatstone's theoretical interest in the working of musical instruments led him to invent several new ones. The *Mechanics' Magazine* of Saturday 8 March 1828 reported that Wheatstone – described as 'the inventor of the kaleidophone' – had devised a new instrument which he called a 'Terpsiphone', from the Greek words meaning 'enjoyment' and 'sound'. The article, which is headed 'New Musical Instrument', is quite brief. In 250 words it says that a column of air will sound if a 'sonorous body' such as a tuning fork is brought near, provided the resonant frequency is correct; Wheatstone 'has devised an instrument, to be constructed upon this principle of the reciprocation of columns of air'. Nothing more is heard of the Terpsiphone, but meanwhile Wheatstone's attention had been drawn to a new class of musical instruments.

The notes produced by musical instruments used in Europe before the early nineteenth century are controlled either by a resonant column of air or by a vibrating mechanical system such as a stretched string. Some instruments, such as the clarinet or the oboe, incorporate a reed, but the reed is not free to vibrate at its own natural frequency. In such instruments the note is determined by the resonant frequency of a column of air in which the reed is placed. The first European musical instrument in which the notes were determined by the natural vibrations of reeds was the German Mund Harmonica, or mouth organ, which was introduced about 1825. It was inspired by instruments brought from China late in the eighteenth century. The accordion, developed in Austria, was essentially a mouth organ fitted with bellows and keys.

Wheatstone's principal musical inventions, the symphonium and the concertina, were developments from the mouth organ and accordion. The difference between the two was that the symphonium was blown by mouth and the concertina by bellows. Both instruments were described in Wheatstone's patent of 19 June 1829, though the word 'concertina' was not used. The main feature of the invention, as described in the patent specification, was the layout of the keys. Other keyed instruments had their keys arranged in a line, but Wheatstone's invention was, in the words of his patent specification:

the employment of two parallel rows of finger studs on each end or side of the instruments fitted with keys to terminate the ends of the levers of the keys, and the so placing them with respect to their distances and positions as that they may, singly, be progressively and alternately touched or pressed down by the first and second fingers of each hand, without the fingers interfering with the adjacent studs, and yet be placed so near together as that any two adjacent studs may be simultaneously pressed down, when required, by the same finger, the peculiarity and novelty of this arrangement consisting in this, that as in the ordinary keyed wind musical instruments the fingering is effected by the motion sideways of the hands and fingers, in this new arrangement that mode of fingering is rendered entirely inapplicable, and a motion which had not hitherto been employed is rendered available, namely, the ascending and descending motions of the fingers before described. This mode of arranging the studs enables me to bring the keys much nearer together than has hitherto been done in any other instrument of a similar nature, and thereby to construct such instruments of greater portability.

In brief, there were two lines of finger studs of such size and spacing that one finger could press a single stud or two adjacent studs in the same line. The notes were arranged so that pressing any two adjacent studs produced a chord, and the instruments were smaller and more portable than those with the keys arranged in a single line.

The symphonium (see Figure 4.1) and concertina were described by Faraday in a discourse given at the Royal Institution on 21 May 1830, under the title 'On

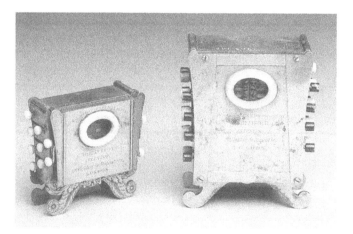

Figure 4.1   The Symphonium.

the Application of a New Principle in the Construction of Musical Instruments'. But the 'new principle' with which Faraday was concerned was not the arrangement of the keys, but the free reed principle:

now so well known for its popularity in the æolina (or mund-harmonica), where a spring of metal being fixed by one end, in an aperture which it nearly fills, is thrown into vibration by the breath or any other soft current of air passing by it, and produces musical sound.

Faraday began the discourse by giving the general laws of vibrations of rods and springs, partly illustrating them with 'an instrument called a tonometer invented by Mr Wheatstone'. (The present writer is not aware of any other attribution of the tonometer to Wheatstone, but in Wheatstone's first lecture on Sound at King's College he mentioned Galileo's demonstration that the time of vibration of an elastic cord is proportional to its length. In the course of his research Wheatstone might well have constructed an instrument in which a cord was kept in constant tension, probably by a weight, and a measured length of the cord used to produce a note of calculable pitch.) During the discourse the symphonium was demonstrated by a Mr Godbé. A number of other reed instruments were also demonstrated and described, including the Tshing, or Chinese Organ, which employed reeds with associated resonating columns of air.

On 11 June 1830 Faraday gave another discourse on the vibration of rods and strings. He described the kaleidophone, an account of which had already been published, and ended by referring to a new mode of counting vibrations and other rapid motions by eye, which was currently being investigated. This was Wheatstone's revolving mirror, which was used in his measurement of the velocity of electricity and will be considered in Chapter 6. There is nothing in the lecture reports to suggest that Wheatstone experimented with different metals for the reeds, though he probably would have done. 'Wheatstone' musical instruments have been made with reeds of different metals, but many of these instruments date from after the end of Charles Wheatstone's active association with the firm.

In July 1836 Wheatstone obtained another patent for musical instruments. It was obtained jointly with John Green, a musical instrument maker of Soho Square, but William Wheatstone was not involved. This patent related to musical instruments in which 'strings, wires or springs' were set in vibration by a current of air directed upon a limited portion only of the length of the string, wire or spring, and also instruments in which the string, wire or spring was struck mechanically as well as blown when the corresponding key was pressed. A further feature was an instrument with its wind-chest divided into two compartments provided with valves so arranged that the currents of air were produced in the same direction whether the bellows were being drawn out or being pressed in.

The firm had staff living in, though some manufacturing work was done by sub-contractors working in their own premises. One former sub-contractor, George Jones, left a manuscript account of his recollections, which begins:

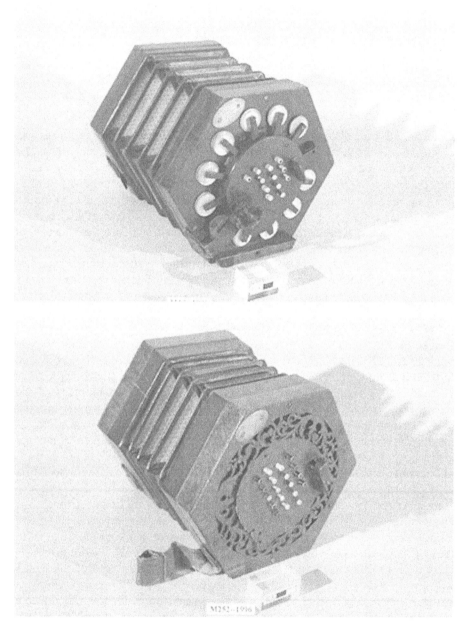

Figure 4.2a   Concertinas in the Wayne Concertina Collection. Two very early Wheatstone concertinas. Serial number 32 (above) has external pallets which are raised when the appropriate buttons are pressed to admit air to the reeds. Most concertinas, including serial number 165 (below) have the pallets inside the instrument. Both instruments are 150 mm across the ends.

Source: By courtesy of the Horniman Museum and Gardens.

Figure 4.2b   *Concertinas in the Wayne Concertina Collection. Bass concertina prototype (above), with no serial number. The rear end has been taken off to show the reeds which, since this is a bass instrument, are fairly large. Concertina serial number 10,660 (below) shows the developed form of the instrument about 1860.*

Source: By courtesy of the Horniman Museum and Gardens.

I was born in 1832, and commenced working in 1844 for a Mr. Austin, who made complete concertina 'pans' for Mr. Wheatstone. The work was all done by hand as 'outwork', and delivered to Wheatstone's for further assembly. Tops, frames and cases were also made as outwork . . . Metal work was made at the factory, where finishing and tuning also took place.[3]

In February 1844 Wheatstone was granted another patent for 'Improvements on the concertina'. There were eight improvements, all concerned with detailed features of the concertina and its construction. None are of much interest considered individually, but taken together they show that Wheatstone was taking a close interest in the technical side of the business at that period. The surviving records of the firm show a steady growth in the sale of concertinas and also list some of the employees. In 1847 a Mr Nickolds and his two sons were engaged as tool makers at 20 Conduit Street. The same year Wheatstone met Louis Lachenal, a Swiss engineer, who was engaged by Wheatstone and rose to take charge of concertina manufacture.

The symphonium never proved very popular. Probably only about 200 were made, and only a few survive. Concertina sales were slow in the early years, only reaching 100 a year in 1844. Serial number 10,000 was reached in 1857. Sales reached 19,000 by the time of Wheatstone's death in 1875, and 23,000 by the end of the century. By that time the concertina was attracting numerous virtuoso performers and composers, as well as amateur musicians. Throughout the 1850s there were many London concerts by The Concertina Quartet. This group was composed of Giulio Regondi, who achieved greater fame as a guitarist, Richard Blagrove, a viola player who took up the concertina as an extra instrument, A.B. Sedgwick, a teacher who produced tutor books and many arrangements for the concertina, and George Case, a player who for some years also had his own concertina factory in London. Blagrove performed for royalty at Windsor. Case gave a concert in May 1851 which, according to the *Musical Times*, was principally remarkable for Rossini's Overture to William Tell, arranged for 12 concertinas with the violoncello solo played on the bass concertina.

Arthur Balfour, British Prime Minister 1902–1905, was an enthusiastic concertina player, and the explorers Shackleton and Livingstone both had one. Concertinas reached the height of their popularity early in the twentieth century when they were often found in working-class 'concertina bands' in the mill towns of the north of England. They were used extensively in Salvation Army bands early in the twentieth century, and they enjoyed a revival in the 1960s and 1970s but are seldom used today for serious music. The Concertina Museum collected by Neil Wayne dates from that time. This collection of over 150 Wheatstone concertinas and related instruments was originally exhibited in Derbyshire but was acquired by the Horniman Museum in south-east London, where it will be featuring in a new gallery opening in 2002.[4]

After Charles Wheatstone had left Conduit Street, William obtained patents for musical instruments, but the brothers never had a joint patent and Charles was to have only one further musical patent. That was granted in 1872, ten years after William's death, when Charles was again taking an active interest in the business. Wheatstone did not leave any written record of his further work on the concertina, but the surviving items show that he tried out many variations of the instrument, including different shapes of reeds and different metals for the reeds. The 1872 patent was obtained jointly with J.M.A. Stroh, who was then working for Wheatstone on telegraphs. It related to reed instruments (harmoniums and concertinas) in which the length of a single reed could be varied to change its note. Instead of having a single reed for each note this instrument had just one reed whose length was varied by a pair of pinch rollers moving along it. When a button was pressed a complex system of levers moved the rollers to adjust the reed to the appropriate length. It was not a success.

After Wheatstone's death the firm continued for many years, being eventually absorbed in Boosey and Hawkes and finally closing in the 1970s.

Teaching musical theory

Wheatstone was always ready to explain scientific facts and new discoveries, provided he could do so in private conversation or in writing. His extreme shyness made it almost impossible for him to address a group of people. Wheatstone's academic interests were not limited to scientific matters. An interest in the history of science is seen in numerous references throughout his published work and he was good linguist. He had an interest in education which probably began in the Lancastrian school, where he would have played his part in teaching younger boys, and he was interested in teaching long before taking a College post. There is no evidence that he had music pupils, but his father certainly did and he would have been aware of the problems involved in teaching musical theory. In 1824 he published the Harmonic Diagram, a teaching aid designed, in his own words, to diminish 'the difficulty attending the acquirement of Musical Theory'. It consists of a disc of card ten centimetres in diameter pivoted at its centre on a larger piece of stiff card. In the pamphlet accompanying the Diagram Wheatstone was at pains to stress that it was not a substitute for any of the 'excellent theoretical works now published', but a useful accompaniment.

For the musically-inclined reader Wheatstone's Harmonic Diagram is reproduced as Figure 4.3. It consists of a card with a rotatable disc fixed by a pin. The circumference represents a perfect octave. If the 'Lyre' index on the edge of the rotatable disc is set to a specific note on the lettered outer circle then the notes of the major diatonic, chromatic and enharmonic scales may be read off, and the key signature may be seen through the aperture in the disc. The seven notes of the diatonic scale are indicated by the thick lines on the disc and it can be seen that the seven intervals consist of five larger intervals (tones) and two smaller intervals (major semitones). The chromatic scale is formed from the diatonic by dividing each of the five tones into a major semitone and a minor semitone. These divisions are indicated by dotted lines on the disc. The enharmonic scale is formed from the chromatic by further subdivision; each major semitone is divided into a minor semitone and a diesis, thus adding seven additional notes

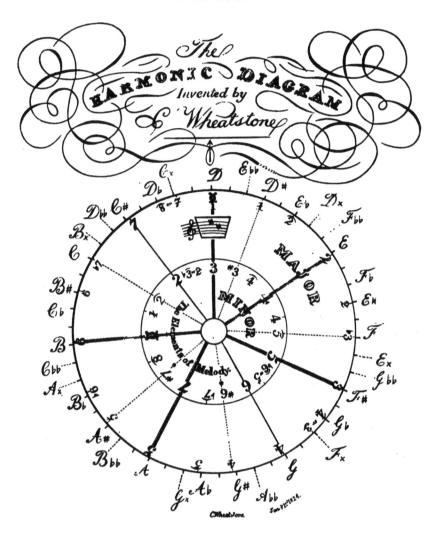

Figure 4.3   Wheatstone's Harmonic Diagram.

which are shown as marks within the major semitones. The five notes of the diatonic scale (marked with double dotted lines) form the 'national' scale, sometimes known as the 'Scottish' scale because traditional Scottish melodies use only these five notes. (The national scale corresponds to the five black keys on the piano.) The large numbers on the disc show the musical interval between each note and the key-note. If a minor scale rather than a major is required then the inner circle of numbers on the disc is used.

The Harmonic Diagram may be used to determine the notes of a scale in any key, and to determine the relationship between major and minor scales. A tune may be transposed between one key and another by setting the index to the key-note of the first key and noting the interval from the key-note of each note in the tune; the index is then moved to the new key-note and the notes corresponding to the intervals can be read off.

There is no evidence that the Harmonic Diagram was a success as a teaching aid, though it was reprinted at least once and two different printings exist which differ very slightly in layout. The Harmonic Diagram is not mentioned in the 32-page 'Catalogue of Music for the Concertina, Harmonium etc.' published by Wheatstone & Co in the 1860s. The catalogue includes music text books and portraits of musicians, so would presumably have included the Diagram if it had been available then. While the Diagram may be an amusing gadget for those who already appreciate the theory of scales, it seems unnecessarily complicated for the beginner. The student attempting to master the theory for the first time would do better with a Diagram showing only the diatonic and chromatic scales, and their division into tones and major and minor semitones. Surprisingly, neither the Diagram nor the pamphlet make mention of the equal temperament scale in which all the semitones are equalized. This scale, which is now universally used on keyboard instruments, had been adopted for most purposes before Wheatstone's time.

The wave machine

Another teaching aid was Wheatstone's wave machine. His experiments with Chladni plates and the kaleidophone, described in the previous chapter, had helped to make the vibrations visible. He wanted to demonstrate the behaviour of light waves, which he envisaged as vibrations in the ether. Experiments related to this were mentioned in letters to Herschel in 1825 and to Faraday in 1831. It was not until the late 1840s, however, that Wheatstone had an early version of what is now called 'Wheatstone's wave machine' (see Figure 4.4). Two visitors to his laboratory saw it: Julius Plücker in 1848 and the Italian P.A. Secchi in 1849.

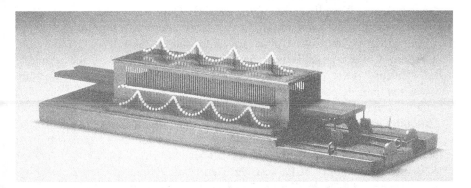

Figure 4.4   Wheatstone wave machine, in the University of Pavia, Italy.

Source: Photo courtesy of Dr Paolo Brenni.

Secchi understood it sufficiently to have a wave machine made when he returned to Rome.[5]

The wave machine was essentially a rectangular brass box with vertical slots. In each slot was a loose element carrying two horizontal wires and a vertical wire, each with a white reflective bead on its end. A number of wave-shaped strips of wood were provided and these could be pushed in grooves through the box making the loose elements move up and down or sideways in ways determined by the shape of the strips. The white beads then moved in a wave motion. By using different strips and having the strips fixed in different relative positions it was possible to demonstrate polarization and interference effects in waves.

A number of Wheatstone wave machines were made and demonstrated in scientific exhibitions in Europe and North America, and several dozen survive. Around 1870 the London instrument makers, Elliott Brothers, were offering the machine for £25, and in 1889 the catalogue of the Paris instrument maker Rudolph Kœnig had two versions, one for 1,000 francs and a smaller model for 600 francs. Kœnig also supplied an iron stand for the wave machine for an extra 150 francs.

In about 1851 the Wheatstones had a family portrait taken by the photographer Antoine Claudet at his studio in London. The family, then Charles, his wife Emma and their first three children are grouped around a table with one of Wheatstone's inventions on the table. The photograph was a stereoscopic pair, but he did not choose to show a stereoscope. The object on the table is a wave machine.

Notes

1 See 'Professor Wheatstone's Case' in the arbitration papers, quoted in Chapter 10 below.
2 Edwin S. Grosvenor and Morgan Wesson, *Alexander Graham Bell, the man who invented the telephone*, 1996, p. 17.
3 George Jones, 'Recollections of the English Concertina Trade', *Free Reed (The Concertina Newsletter)*, 16, Derby, November 1973, pp. 14–20.
4 I am grateful to Neil Wayne for information about the Concertina Museum Collection and the concertina business. Much of the available information is to be found on the Free Reed Magazine website: www.freedmus.demon.co.uk.
5 Julian Holland, 'Charles Wheatstone and the Representation of Waves', Rittenhouse, 13, 1999, pp. 86–106. V.K. Chew, manuscript translation of P.A. Secchi, 'Sopra una nuova macchina per rappresentare i moti vibratori delle ondulazioni luminose inventata dal Sig. Wheatstone', *Corrispondenza Scientifica di Roma*, 1850. Paolo Brenni, entry on Wheatstone's wave machine in *Gli strumenti fisica dell'Istituo Technico Toscano – Ottica*, Florence, 1995.

Chapter 5

The stereoscope

The stereoscope was one of the most popular scientific toys of the nineteenth century.[1] Only the kaleidoscope was sold in greater numbers. The stereoscope produces the illusion of a three-dimensional scene or object from two slightly different flat pictures which are viewed through the apparatus in such a way that each eye sees only one of the pictures. Because the two eyes are spaced a short distance apart the image of a solid object projected on one retina differs slightly from the image of the same object on the other retina. Wheatstone was the first person to realize that the brain makes use of these differences to determine not only the relative distances of different objects but also the relative distances of different parts of one object and so give an impression of relief. Before his discovery it was assumed by writers on optics that the differences between the images of a single object on the two retinae were so small as to be negligible.

Binocular vision had attracted the interest of scientists for centuries. In his *Treatise on Optics* Euclid (third century BC) showed that a person looking at a sphere saw either more than half, exactly half, or less than half of the circumference depending on whether the diameter of the sphere was less than, equal to, or more than the distance between the observer's eyes. It is clear that Euclid appreciated that the two eyes saw a slightly different view of the sphere: the left eye saw more on the left and the right eye saw more on the right. There is nothing, however, to suggest that Euclid understood the stereoscopic effect achieved with binocular vision. The physician Galen (second century AD), in his *On the use of the different parts of the Human Body*, made the point that a person standing near a column and observing first with one eye and then with the other will see different portions of the background. The left eye, for example, will see a part of the background to the left of the column which is hidden from the right eye. Leonardo da Vinci (1452–1519) made the same point in his *Trattato della Pittura* when discussing why a plane painting can never show relief in the same way as a solid object. In each of these examples the sphere or column, being circular, presents a similar view to each eye. Galen was concerned only to find a reason for having two eyes. He concluded that it was to increase the amount seen. If Leonardo had considered a square column, and if he had concentrated his attention on the column itself and not the position the column occupied relative to the background, he might have considered the significance of the two eyes seeing different views of the column. This was the starting point of Wheatstone's work which led to the stereoscope.

The earliest published allusion to the stereoscope is in the third edition of Herbert Mayo's *Outlines of Human Physiology*, published in 1833, the year before Wheatstone's appointment to King's College. Mayo was the first professor of anatomy and physiology at the College, from which he resigned in 1836. Referring to 'a paper Mr. Wheatstone is about to publish', he wrote:

One of the most remarkable results of Mr. Wheatstone's investigations respecting binocular vision is the following. A solid object being placed so as to be regarded by both eyes, projects a different perspective figure on each retina; now if these two perspectives be accurately copied on paper, and presented one to each eye so as to fall on corresponding parts, the original solid figure will be apparently reproduced in such a manner that no effort of the imagination can make it appear as a representation on a plane surface. This and numerous other experiments explain the cause of the inadequacy of painting to represent the relief of objects, and indicate a means of representing external nature with more truth and fidelity than have yet been obtained.

Mayo's reference is by no means clear, and it would not be surprising if his first readers passed this paragraph by without understanding what Wheatstone had done. The reader who knows the principle of stereoscopy will appreciate that that principle is in fact to be found in Mayo's elaborate language.

Wheatstone was interested in vision at least as early as 1826 when he wrote to Örsted that he intended soon to publish 'Some remarkable phenomena of Vision which have fallen under my observation'. The first public appearance of the stereoscope was before the Royal Society in 1838 when Wheatstone presented his paper 'Contributions to the Physiology of Vision – Part the First. On some remarkable, and hitherto unobserved, Phenomena of Binocular Vision'. He had previously been responsible for two papers with the title 'Contributions to the Physiology of Vision' in the *Journal of the Royal Institution* in October 1830 and May 1831. These were not reporting his own research, but were seen as the start of a series in which 'it is proposed to bring forward those stores of knowledge on this subject which have been hitherto locked up in the repositories of foreign scientific literature'. This 'Contributions to the Physiology of Vision, No. 1' was largely concerned with the work of Dr J. Purkinje, the Professor of Physiology at Breslau, who had published an 'Essay on the Subjective Phenomena of Vision' at Prague in 1823. It gave the substance of Purkinje's Essay in a well-written article of about 6,000 words, with only a short editorial comment at the end. 'Contributions . . . No. 2' was much shorter, about 1,300 words, and was a series of comments on the first article, not a translation of foreign work. The *Journal of the Royal Institution* ceased publication shortly afterwards, and no further 'Contributions to the Physiology of Vision' were published.

Wheatstone's paper to the Royal Society in 1838 describing the stereoscope (see Figure 5.1) was one of his longest papers – over 12,000 words. He claimed as a new fact in the theory of vision the observation that two dissimilar pictures were projected on the two retinae when a single object was viewed. He then asked: 'What would be the visual effect of simultaneously presenting to each eye, instead of the object itself, its projection on a plane surface as it appears to the eye?' To answer his own question he described apparatus which he called the 'stereoscope' for trying the experiment. The original stereoscope which Wheatstone demonstrated to the Royal Society is now in the Science Museum (see Figure 5.2). In his paper he made the point that it was not essential to have any special apparatus for viewing a pair of stereoscopic pictures provided that the width of the pictures was less than the distance between the viewer's eyes. It is

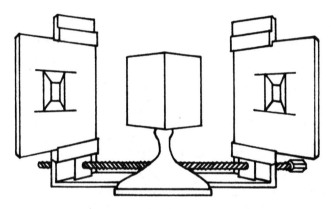

Figure 5.1   Earliest form of Wheatstone's stereoscope (from his paper of 1838).

Figure 5.2   Reflecting stereoscope by Wheatstone (now in the Science Museum).

nevertheless helpful to have some mechanical aid such as a pair of tubes to look through so that the left eye is not distracted by a peripheral view of the right eye's picture, and vice versa.

There was no mention of the use of prisms instead of mirrors in the 1838 paper, although Wheatstone had used prisms by then. Another paper on the stereoscope, in 1852, did mention the use of prisms. This later paper was a continuation, after 14 years, of the earlier one. It included a description of a neatly designed, folding stereoscope which collapsed into a box (see Figure 5.3). There are instruments of this pattern in the Science Museum.

Wheatstone's stereoscope was received enthusiastically at the British Association meetings at Newcastle in September 1838, but it was a purely scientific interest. Popular and commercial interest had to await the development of photography. The 1838 paper was accompanied by a number of stereoscopic pairs of line drawings, and some of these are reproduced in Figure 5.4. The reader may be able to see these drawings three-dimensionally without any apparatus, but it will be easier if a piece of card or paper is held vertically on the centre line of the page so that each eye sees the drawings on one side only. It is best to begin by focusing the eyes on a distant point behind the page, and then bring the focus forward.

The invention of photography made it possible for stereoscopic pairs of pictures to be produced with the necessary accuracy. Wheatstone quickly had stereoscopic photographs prepared. In the 1852 paper he wrote:

At the date of the publication of my experiments on binocular vision, the brilliant photographic discoveries of Talbot, Niepce, and Daguerre had not been announced to the world. To illustrate the phenomena of the stereoscope I could therefore, at that time,

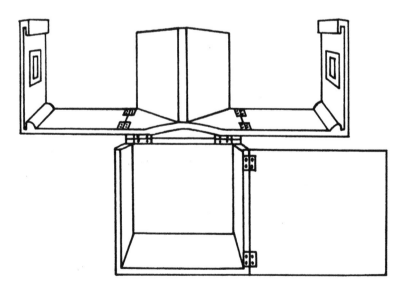

Figure 5.3   Wheatstone's folding stereoscope (from his paper of 1852).

Figure 5.4   Stereoscopic pairs of drawings (from Wheatstone's paper of 1838).

only employ drawings made by the hands of an artist. Mere outline figures, or even shaded perspective drawings of simple objects, do not present much difficulty; but it is evidently impossible for the most accurate and accomplished artist to delineate, by the sole aid of his eye, the two projections necessary to form the stereoscopic relief of objects as they exist in nature with their delicate differences of outline, light, and shade. What the hand of the artist was unable to accomplish, the chemical action of light, directed by the camera, has enabled us to effect.

It was at the beginning of 1839, about six months after the appearance of my memoir in the Philosophical Transactions, that the photographic art became known, and soon after, at my request, Mr. Talbot, the inventor, and Mr. Collen (one of the first cultivators of the art) obligingly prepared for me stereoscopic Talbotypes of full-sized statues, buildings, and even portraits of living persons. M. Quetelet, to whom I communicated this application and sent specimens, made mention of it in the Bulletins of the Brussels Academy of October 1841. To M. Fizeau and M. Claudet I was indebted for the first Daguerreotypes executed for the stereoscope. The beautiful stereoscopic representations of statuary, architecture, machinery, natural history specimens, portraits of living persons, single and in groups, &c., which have recently been produced by M. Soleil and M. Claudet, are now too well known to the public to need more than a slight reference to them . . .

For obtaining binocular photographic portraits, it has been found advantageous to employ, simultaneously, two cameras fixed at the proper angular positions.

Later Henry Collen himself gave an account of how he made a stereoscopic pair of photographs of Babbage at Wheatstone's request and 'under his direction' in August 1841. Wheatstone and Antoine Claudet worked together for a time on the commercial application of photography to the stereoscope, using the daguerreotype process, and in 1839 Wheatstone told Herschel in a letter that he and a Mr Brockledon were visiting a Mr Bauer at Kew Green to see specimens of Niepce's photographic process.

Wheatstone does not appear to have taken any further active interest in the commercial exploitation of the stereoscope. Sir David Brewster devised an alternative form of stereoscope using lenses in 1849. Brewster preferred the use of two half lenses with their thin edges adjacent. It was Brewster's form of stereoscope which was adopted almost exclusively when in the 1850s and later vast numbers of stereoscopes and stereoscopic pictures were sold. Many patents were obtained for new designs of stereoscope, and Table 5.1 illustrates the rise and fall of an industry.[2]

In 1856 there was an exchange of correspondence between Brewster and Wheatstone in *The Times*. The occasion was the announcement of a new

Table 5.1 British Patents granted for stereoscopes

1853	2	1858	11	1863	3
1854	1	1859	11	1864	2
1855	2	1860	7	1865	2
1856	9	1861	0	1866	2
1857	16	1862	7	1867	0

stereoscope designed by the Frenchman Faye. This was simply a piece of card with two holes spaced apart at the same distance as the user's eyes. When held between the eyes and a pair of stereoscopic pictures it served to guide the eyes and prevent each eye seeing the wrong picture. The purpose of Brewster's first letter, which was published anonymously, over the signature 'A', was to claim that the stereoscope was invented by Mr James Elliott, a teacher of mathematics in Edinburgh. He 'contrived' it in 1834 but 'did not execute it till 1839'. The claim that Elliott was the inventor of the stereoscope had been put forward by Brewster in his book *The Stereoscope*, published earlier the same year. Elliott's stereoscope consisted of a box about 45 centimetres long. At one end were two dissimilar pictures of a scene as seen by each eye, and there were two eye holes at the other end. The scene had moon, sky and mountain in the background, a cross in the middle distance, and a branch of a tree nearest to the observer.

Wheatstone, apparently unaware of the identity of 'A', wrote to *The Times* pointing out that his discovery of the principle of stereoscopy had been published in 1833 in the book by Mayo and that actual stereoscopes had been described in his own paper of 1838, shown at the British Association that year and reported in the *Athenaeum* and the *Literary Gazette*, and yet:

Sir D. Brewster and your correspondent in accordance with him, represent Mr. Elliott as having conceived the idea of a stereoscope in 1834, and as having realized his conception in 1839.[3]

Brewster wrote again, in his own name this time, saying that he had written anonymously before 'not thinking it of sufficient importance to add my name'. He concedes equal credit for the invention of the stereoscope to Wheatstone and Elliott, but argued that the principle of stereoscopy was known from antiquity: Euclid and Galen, among others, had known it. In his second letter Wheatstone quotes a letter from the instrument maker R. Murray to establish that he had had a stereoscope in 1832:

43 Piccadilly, October 27

Sir, -From an examination of the accounts furnished to you by Mr. Newman, of Regent-st, during the time I was in his establishment, and which were prepared by myself, I am enabled to assign the date of my first knowledge of your stereoscope, both with reflecting mirrors and refracting prisms, to the latter part of 1832.

I am, Sir, Yours faithfully,

H. MURRAY

Professor Wheatstone.

Wheatstone then continued:

The undue prominence given to Mr. Elliott's single experiment may lead some persons to imagine that the results he obtained were at least as perfect as those which I had previously produced; but it appears that he did not proceed so far as to give the representation in relief of any solid body whatever; his attempt, as described by Sir D. Brewster in his recent work, was limited to represent three different flat distances, to either of which the eyes might be converged at will. The name 'stereoscope' is quite inappropriate to an instrument exhibiting this effect alone.

Sir D. Brewster, not content with disputing my right to be considered the inventor of the stereoscope, denies, even if that were to be admitted, my claim to the discovery of the principle on which it is founded. The real fundamental principle of the stereoscope is that clearly stated in my earliest announcement – namely, the apparent reproduction of a solid object by simultaneously presenting its two perspective projections, artificially delineated, one to each eye.

Finally Wheatstone quoted Brewster's own words in a paper of 1843 in the *Transactions of the Royal Society of Edinburgh*:

In prosecuting this subject, my attention has been particularly fixed upon the interesting paper of my distinguished friend Professor Wheatstone, on some remarkable and hitherto unobserved phenomena of binocular vision. It is impossible to overestimate the importance of this paper, or to admire too highly the value and beauty of the leading discovery which it describes – namely, the perception of an object of three dimensions by the union of the two dissimilar pictures formed on the retinae.

Brewster followed with another long letter in which he puts forward another 'inventor', George Maynard of Toronto, who, according to Brewster, wrote 'a protracted article, signed 'Theophilus', and involving a detailed enunciation of binocular phenomena' which was published in a Canadian newspaper. However, that was only in 1836.

Wheatstone's third letter brought to an end the exchanges in *The Times*. Wheatstone's attitude to the correspondence is revealed by a couple of brief extracts from this letter:

I have hitherto avoided entangling myself in the meshes of controversy with so disputatious an antagonist as Sir D. Brewster. I have always thought myself more usefully employed in investigating new facts than in contending respecting errors which time will inevitably correct . . .

I was far from thinking, when answering an anonymous letter in the columns of The Times, that it had emanated from the same source from which had proceeded all the attacks which have with reference to this matter during the last four years been directed against me; but I cannot regret the opportunity which that circumstance has afforded me to correct, in the most efficacious manner, a few of the most prominent of the mis-statements made.

Finally Wheatstone points out again that Brewster had failed to show that any previous writer had stated the principle of stereoscopy.

The Times correspondence ended here. Brewster, however, continued to dispute Wheatstone's claim to the discovery and in 1860 he read a paper to the Photographic Society of Scotland entitled 'Notice respecting the Invention of the Stereoscope in the sixteenth century, and of Binocular Drawings by Jacopo Chimenti da Empoli, a Florentine artist'. Wheatstone appears to have taken no further part in the controversy. The reader who wishes to pursue Brewster's case further is referred to the article by A.T. Gill cited above. Gill quotes with approval Sir John Herschel's remark that Wheatstone was the inventor of the stereoscope, Brewster invented a way of looking at stereoscopic pictures. The concept of using two plane views to produce the illusion of a three-dimensional object is due to Wheatstone. The form of stereoscope most widely adopted in the nineteenth century is due to Brewster.

Stereoscopy enjoyed a revival in the years following the Second World War. Neither Wheatstone's nor Brewster's forms were used then, but the Anaglyph process invented by Duhauron in 1891. In this the two pictures are printed together, one in red and the other in green, and the viewer has to wear spectacles with one red and one green lens. Stereoscopic films, using such spectacles, were demonstrated at the Festival of Britain in 1951, but were not a commercial success. They are still seen occasionally.

The parallax panoramagram process was developed in the early years of the twentieth century. It has the pictures printed in fine interleaved strips and viewed through a grid which ensures that each eye sees only what is intended. In recent years picture postcards have been produced in this way with the grid a plastic sheet incorporated into the card. The great advantages of the system are that the picture can be in colour and the observer needs no special equipment to view the picture in three dimensions.

The second two-thirds of the first paper in which Wheatstone describes the stereoscope, and most of the second paper, are devoted to a variety of observations on aspects of vision. In the first paper he states the fact that if a pair of stereoscopic pictures are interchanged the observer perceives a three-dimensional figure which he calls the 'converse' of the original. Much of the second paper is concerned with the 'pseudoscope', an instrument which enables an observer to see the 'converse' of an actual object. The pseudoscope has prisms so arranged that the right eye sees the left eye's view, and vice versa. Peculiar effects may be observed with the Pseudoscope. The inside of a tea-cup appears as a solid convex body; a bust becomes a hollow mask; a medal becomes the die from which it was struck. Wheatstone notes that it is important to have the object uniformly illuminated because the presence of light and shade has a considerable effect on the perception of relief.

One of the earliest observations which, according to Wheatstone, had drawn his attention to the subject of binocular vision is an effect which may be seen in a metal plate made smooth in a lathe:

When a single candle is brought near such a plate, a line of light appears standing out from it, one half being above, and the other half below the surface; the position and inclination of this line changes with the situation of the light and of the observer, but it always passes through the centre of the plate. On closing the left eye the relief disappears, and the luminous line coincides with one of the diameters of the plate; on closing the right eye the line appears equally in the plane of the surface, but coincides with another diameter; on opening both eyes it instantly starts into relief.

This phenomenon can be seen using a spun aluminium plate such as the bottom of a saucepan and either a candle or an electric light with a small source area. It is necessary for the observer to look almost perpendicularly onto the plate, for the light source to be almost in the plane of the plate, and for the observer, the plate and the light source to be aligned so that the line of light appearing on the plate points towards the light source. Under these conditions (which could easily arise when working by candlelight) the line of light appears to stand out from the plate, half being above the plate and half below. The apparent inclination of the line is such that it points beneath the light source.

Notes

1   The sources for this chapter, apart from Wheatstone's papers, are R.S. Clay, 'The Stereoscope' (Presidential Address) *Transactions of the Optical Society,* **29**, 1928, pp. 14–66. Henry Collen, 'Earliest Stereoscopic Portraits', *Photographic Journal* I, 1854, p. 200. A.T. Gill, 'Early Stereoscopes', *Photographic Journal*, 109, 1969, pp. 546–59, 606–14, 641–51. D. Brewster, 'Description of several new and simple stereoscopes', *Transactions of the Royal Scottish Society of Arts*, 3, 1849, pp. 247–64. D. Brewster, *The Stereoscope*, 1856.
2   A.T. Gill, 'Early Stereoscopes', *Photographic Journal*, 1969.
3   Correspondence in *The Times* 17, 20, 25 and 31 October and 5 and 15 November 1856.

PART 2
ELECTRICITY

Chapter 6

The velocity of electricity

Wheatstone considered that the velocity of electricity would be an important factor in developing an electric telegraph. He knew that sound was carried best in those materials in which it travelled fastest, and he assumed the same would be true for an electric signal in a wire. In the experiments described in this chapter he tried to measure two different things, and he seems to have confused the two in his own mind. First he attempted to measure the speed with which an electric spark passed through the air, but it proved to be so fast that he was unable to make any estimate of the speed. His second experiment, which was more successful, was to measure the speed with which an electric signal travelled in a wire. He arranged a circuit with three spark gaps which were connected through several miles of wire but located side by side, and then measured the time which elapsed between the occurrence of a spark at each spark gap when a current was sent round the circuit. For this measurement he used a high speed, rotating mirror which he had already used in acoustic research.

The motion of a sounding violin or piano string is far too rapid for the unaided eye to follow, but Wheatstone wanted to know how strings and other vibrating bodies moved. He had shown with his Kaleidophone that the path of a luminous or illuminated point in rapid motion appears as a continuous line because of the persistence of vision. It occurred to him that if this motion were combined with another motion of known speed in a direction at right angles to the first then it would be possible, by observing the resultant line, to work out the original motion. He made various acoustic experiments using this idea. He knew that under certain circumstances a gas flame in a glass tube would oscillate, producing a musical note, and later he made an experimental 'organ' in which sound was produced in this way. He found that a rotating mirror enabled him to observe the movement of the flame, and he then considered whether it might also be used to find the direction and velocity of an electric spark. His proposals were first announced by Faraday in the lecture on 11 June 1830, although he only took the subject up actively in early 1833.

His interest may in fact have started much earlier, for as a boy of about 15

Wheatstone had been present at some of Sir Francis Ronalds' experiments with an electric telegraph at Hammersmith. One of these was an attempt to measure the speed of electricity in a wire by observing the time which elapsed between the connection of an electrical machine at one end and the firing of a cannon at the other end. Ronalds said he was unable to observe any time lapse since the cannon seemed to fire at the moment the connection was made.[1]

Wheatstone first set out to measure the velocity of an electric spark passing through the air. His apparatus (Figure 6.1) had an arm rotating 50 times per second. When the ball *f* was within striking distance of the prime conductor of a static electric machine a spark passed between them and also between the balls *g* and *h*, which were ten centimetres apart. With this apparatus Wheatstone hoped to determine the velocity of a spark passing through the air by observing the direction of the spark. He reasoned that if the spark passed between the spheres with a velocity of the same order as the velocity of the moving sphere, then the spark would appear to be deflected from the vertical. No deflection was observed, therefore the spark was moving much more quickly than the moving ball. If the spark took one thousandth of a second to cross the gap, the upper end should be deflected sideways relative to the lower end by the distance travelled by the upper sphere in that time. At 50 revolutions per second this would be one twentieth of the circumference of its path of travel. Such a deflection would easily be seen. Wheatstone therefore concluded that the spark must pass in less than one thousandth of a second.

Wheatstone next tried fixing the spark gap and observing not the spark itself but the image of the spark seen in a rapidly revolving mirror. A small movement applied to the mirror produced a much greater movement of the image of the object than that same movement would produce if applied to the object itself.

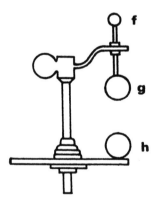

Figure 6.1   *Wheatstone's first experimental attempt to measure the velocity of an electric spark in air. The upper part rotates on the vertical shaft and a spark is drawn between the balls **g** and **h**. If the balls were not moving the spark would appear vertical. Wheatstone reasoned that if the speed of the ball **g** were comparable with the speed of the spark across the gap then the spark would appear inclined.*

Also the mirror and its rotating mechanism constituted an independent instrument which might be employed to investigate a variety of phenomena. With the mirror in a vertical plane and revolving rapidly about a vertical axis (Figure 6.2a) he observed the image of a ten-centimetre vertical spark. The movement of the mirror was not in any way synchronized with the occurrence of the spark, but whenever the mirror happened to be in a position for the image of the spark to be seen, the image was identical with the image seen when the mirror was at rest. He calculated that a time interval of $\frac{1}{72\,000}$ of a second would have been detectable with this device. The duration of the spark was, therefore, less than that time at all points.

With the mirror turned almost perpendicular to the axis (Figure 6.2b), the image of an object such as a candle flame was formed into a circle of light. He used this arrangement to show that when a gas flame was arranged to burn in a glass tube so as to produce a musical note it underwent a regular variation of intensity.

Wheatstone decided to apply the revolving mirror apparatus to the measurement of the velocity with which electricity passed through a conductor. Previous experimenters had attempted to do this by detecting a time interval between discharges across spark gaps at opposite ends of a wire, the ends being brought close together so that the spark gaps might be observed simultaneously. No-one had managed to detect any time interval. Probably the longest circuit employed was the four miles (two miles of wire and two miles of earth return) used by Dr Watson at Shooter's Hill, London, in 1747.

Experiments such as Dr Watson's were based on the assumption that an 'electric fluid' passed from one end of the wire to the other. Wheatstone pointed out that if electricity were two 'fluids' (or opposite disturbances of equilibrium of a single 'fluid', travelling from opposite ends of the wire), then sparks at each end of the wire would occur simultaneously whatever the velocity of electricity. However, if a third spark gap were connected at the centre of the wire, the discharge there would occur later in time than the discharges at the ends of the wire.

Wheatstone conducted these experiments at the Gallery of Practical Science in Adelaide Street, off the Strand. The Adelaide Gallery was set up by a small

 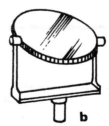

Figure 6.2   The rotating mirror, for viewing sparks or any high speed motion.

group of wealthy men to display machinery and scientific discoveries and to educate the public. There was a long tank for demonstrating ship models, and part of the tank was deep enough to take a diving bell in which visitors could go down at specified times. Unfortunately the Gallery did not attract the public in sufficient numbers, and so did not last very long.

Wheatstone set up half a mile of insulated copper wire. Spark gaps, connected at each end and at the centre of the circuit, were arranged side by side on a board so as to be viewed simultaneously by the mirror which revolved 800 times per second. The apparatus is shown in Figures 6.3 and 6.4. The discharges were synchronized with the mirror by a spark gap switch on the mirror axis and the speed of the mirror was regulated by observing the note of a siren on its shaft or the sound made by a part of the rotating system striking a fixed slip of paper. Later a counter was fitted, but the load which this imposed slowed the apparatus from a maximum of 800 revolutions per second to about 600.

With this apparatus he made two discoveries. The spark lasted for a measurable time, the longest time he observed being $\frac{1}{24\,000}$ of a second. The spark at the middle gap in the circuit occurred later than the sparks at the end gaps, which were simultaneous. The deviation between the central spark and the outer sparks as viewed in the mirror of his apparatus was not more than one half of one degree. This figure corresponds to a velocity of electricity of 288,000 miles per second – about one-and-a-half times the velocity of light, which was known at the time.

Figure 6.3   *Wheatstone's apparatus for measuring the velocity of electricity. The mirror is rotated at high speed and the spark balls on the same axis were intended to complete the circuit from the Leyden jar so that the spark always occurred at the same angular position of the mirror.*
(Redrawn from Wheatstone's drawing and description.)

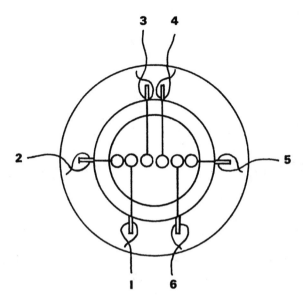

Figure 6.4   *Wheatstone's spark board with three pairs of spark gaps. The spark gaps 1, 2 and 5, 6 were at opposite ends of a long wire, and spark gaps 3, 4 were in the middle. When a Leyden jar was discharged through the wire a spark occurred at all three gaps. The sparks were viewed through a mirror rotating about a horizontal axis. If the sparks were simultaneous then they would appear in the same horizontal position. If there were a time difference between the sparks the mirror would have moved and the sparks would not appear in the same line. In this way Wheatstone showed that the sparks at the ends were simultaneous but the spark in the middle occurred later.*

(Redrawn from Wheatstone's drawing and description.)

These results were published in his paper, 'An Account of some Experiments to measure the Velocity of Electricity and the Duration of Electric Light', submitted to the Royal Society on 14 July 1834. In an accompanying letter Wheatstone wrote:

I am sorry that I have given you so much trouble respecting my paper, which I now send. Fig. 6 is the only omission and this will be ready in a day or two. I have been much divided between the desire of rendering the subject more complete and the fear of delaying its publication another year; but if on perusal you think that it will be better to continue the experiments further before the paper is published I will willingly withdraw it for the present.[2]

Wheatstone was uncertain whether to publish the paper at that time or continue his experiments first. He may have published because he was under pressure to do so, although he would have preferred to have better results to report. His work had been announced prematurely and inaccurately by W.H. Fox Talbot (1800–1877) in the *Philosophical Magazine*. Talbot, who had not yet made his discoveries in photography, had just been elected Liberal Member of Parliament

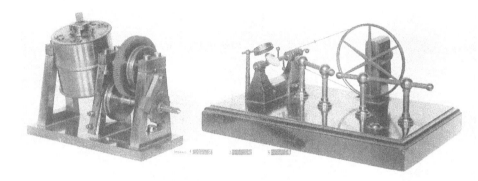

Figure 6.5   Wheatstone's revolving mirror apparatus, now in the Science Museum.

for Chippenham and had been a Fellow of the Royal Society for two years. He wrote a series of articles for the *Philosophical Magazine* under the title 'Proposed Philosophical Experiments', the article on the velocity of electricity appearing in August 1833:

Some ingenious experiments have lately been made by Mr. Wheatstone, with a view to determine the velocity of the electric-spark passing through air by means of a revolving mirror. But has it ever been shown with certainty that the passage of electricity, even through a conducting body, is performed in a space of time so short as to be absolutely inappreciable? An experiment is upon record in which the spark was sent through seven miles of iron wire, which it is said to have traversed in an instant. But it may fairly be presumed that the philosophers who made this observation, could not answer for an interval of time smaller than the tenth part of a second.

Now, as the revolving mirror gives us the power of increasing the accuracy of observation at least a thousand-fold, I will suggest a method of applying it to determine this question.

Let the greatest length of wire that can be procured, be disposed so that the two extremities are brought very nearly together. Let one end of the wire receive the spark from the machine, and the other end give it out again to any body which communicates with the earth.

If the flashes of electric light, entering the wire, and leaving it after traversing its whole length, appear simultaneously to the eye, take a mirror mounted on a revolving axis, and place it in such a position that (the mirror being at rest) the images of the two sparks coincide, or are superimposed one upon the other.

This being effected, let the observation be made through a fixed tube; then if the mirror be made to revolve with great speed, if any separation of the combined spark into two take place, it will be a proof of the existence of an interval of time between them.

The necessity of the tube is apparent; for if the eye were directed to other points of the revolving mirror, the two images would appear separate from the mere effect of perspective.[3]

Wheatstone wrote a long letter to the Editor, Sir David Brewster, which was published the following month:

Dear Sir,

In the last number of the Philosophical Magazine is inserted an article entitled 'Proposed Philosophical Experiments by H.F. Talbot, Esq. MP, FRS'. Upon a part of this communication, headed 'On the velocity of Electricity', I feel it necessary to make the following observations.

My experiments were not made solely with a view to determining the velocity of the electric spark passing through air as Mr. Talbot has inferred, but, from the first, were intended to extend to the passage of electricity through solid conductors. In fact, the very first experiment of the kind which I made, and which was shown to the members of the Royal Institution on the evening my investigations on this subject were first made public by my friend Dr. Faraday, was one by which I endeavoured to show the deviation from a vertical line of two sparks simultaneously visible at the opposite ends of a metallic conductor. Beyond this, I have for several months past communicated very generally to my scientific acquaintance the details of an experiment upon a larger scale, which I hope soon to have an opportunity of executing, and which will detect and measure the velocity of electricity in its passage through a metallic conductor, though the rapidity of its transmission may exceed that of light – this I have proposed to effect, by increasing, in certain proportions, 1st, the velocity of the revolving mirror; 2ndly, the length of the conducting wire; and 3rdly, the accuracy of observing the deviation of the sparks from a vertical line. If I succeed in this point, it is obvious that we shall possess a means of directly measuring the relative conducting powers of metal, and of ascertaining numerous particulars respecting ordinary electricity which we at present have no means of determining.

Intending in the ensuing session to submit to the Royal Society all the results I have obtained, in reference to a new optical means of measuring rapid motions, minute intervals of time, and feeble intensities of light, I have hitherto refrained from publishing any incomplete statement of them; but I regret that this delay should have occasioned my experiments to be so far misunderstood, that one of the earliest which suggested itself to me, and which I have always considered to be of primary importance in the series, should be proposed elsewhere, several months afterwards, as an experiment yet to be tried, and be represented also as having entirely escaped my attention.

I remain, Dear Sir, yours &c
C. Wheatstone
Conduit Street, Hanover Square
August 2, 1833

After that exchange of correspondence Wheatstone may have felt obliged to publish the work he had done so far, even though the results were incomplete, as he made clear in the paper:

The preceding experiments have been directed rather to detect elongations and deviations than to measure them, I am not prepared to state the results with numerical accuracy. I shall endeavour to supply this deficiency in further investigations.

Despite the careful wording of Wheatstone's paper, the figure of 288,000 miles per second, which he used when discussing the greatest speed his apparatus could detect, was later seized upon by some telegraph promoters to advertise the speed of their service. A number of later writers have made the same mistake.

E.T. Whittaker, in his *History of the Theories of Aether and Electricity*, wrote that Wheatstone 'by examining in a revolving mirror sparks formed at the extremities of a circuit, found the velocity of electricity in a copper wire to be about one and a half times the velocity of light'.[4] Wheatstone never, in fact, published further figures from these experiments. The original paper was republished in German in 1835 and in French in 1842, although it is of course possible that this was done without Wheatstone's knowledge.[5]

In 1834, the year that Wheatstone published the experimental work just described, he was appointed Professor of Experimental Philosophy at King's College London. That appointment will be considered more fully in the next chapter. What is relevant here is that in February 1836 the College Council gave Wheatstone permission to lay down 'a series of iron and copper wires in the vaults of the College for the purpose of trying some experiments in electricity on account and at the expense of the Royal Society'.[6]

Wheatstone obtained a grant of £50 from the Royal Society for this work. In the summer of 1835 he set out his experimental proposals in two long letters to the Secretary of the Royal Society, P.M. Roget (1779–1869), better known today for his *Thesaurus of English Words and Phrases*. A Sub-Committee of three – Messrs Peacock, Whewell and Roget – was appointed by the Council to consider the proposals and Wheatstone's request for a grant. They reported their formal approval in March 1836 – a month after the College Council had given its consent to the installation of the wires.

The first part of Wheatstone's proposals was simply to use a longer circuit than hitherto. He wrote to Roget:

The most extensive circuit I have yet employed has been only half a mile in length, but as the deviations of the reflected sparks will be more considerable in proportion to the length of the wire, I propose to form a circuit of four miles of copper wire $\frac{1}{16}$ of an inch in thickness. The wire to be properly insulated, and stretched along the basement gallery of King's College, and each half mile capable of being connected and disjoined as occasion may require. Four terminations of the wire to be brought along the external wall into the lecture room so that comparison may be made between the time occupied by the passage of electricity through 1 mile and 3 miles, $\frac{1}{2}$ a mile and $3\frac{1}{2}$ miles etc.

Four miles of iron wire of the same thickness to be suspended for the purpose of comparing the velocity of electricity in two different metals. This wire may be occasionally connected with the copper to form a continuous circuit eight miles in extent.[7]

The second part of the proposal was a refinement of the observational technique. In his first experiments at the Adelaide Gallery he used three spark gaps mounted in a line on a wooden board (fig. 6.4). This piece of apparatus is now in the Science Museum. In writing to Roget he explained that although it was easy to see that the middle spark occurred later in time than the outer sparks it was virtually impossible to measure the difference, because it was not possible to arrange for successive sparks to occur at exactly the same angular position of the mirror:

Notwithstanding all the precautions that can be taken it seems nearly impossible to obtain, on each repetition of the experiments, the reflected images of the sparks while the mirror is precisely in the same position. Though the deviations can be well seen by the eye, they cannot, therefore, be measured by any micrometric apparatus. This difficulty may be avoided, and the deviations measured within very narrow limits by changing the positions of the sparks themselves. Thus, when the sparks on the spark board are in the same horizontal line, and the mirror of the measuring instrument is moving with great rapidity, the lines are seen in the mirror, having their terminations thus

now the places of the sparks may be so adjusted that, at a given velocity of the mirror, the ends of the lines may appear in the same right line; the deviations of the places of the sparks necessary to produce this adjustment may be accurately measured, and the angular deviations of the ends of the lines in the mirror where the sparks are in the same horizontal line, may thence easily be calculated. Two spark boards with these adjustments measured on accurate scales will be required.

Wheatstone said that for a cost not exceeding £50 he could obtain this proposed apparatus and also 'a few instruments for measuring with great accuracy the different intensities of electrical discharges, etc.' and that he would then be able to ascertain the velocity of electricity in a conducting wire with greater accuracy than before, the comparative velocities of electricity in iron and copper wires, whether electricity travels with uniform velocity or with a rapidity uniformly accelerating, as several recent writers on electricity had assumed, and whether all kinds of electricity travel with the same velocity or whether their rapidity is influenced by variations of tension or circumstances.

No further experimental results were published, but the experiment was demonstrated at King's College in June 1836 and again in June 1837 as the climax of a series of evening lectures. One account of this says that he employed a circuit of $\frac{1}{16}$ inch diameter copper wire four miles long. There is no evidence that he ever actually used a circuit of *iron* wire. He continued, however, with the investigations and was prepared in private conversation to give figures for the velocity of electricity. Edward Copleston, the Bishop of Llandaff and a leading figure in the College Council, recorded in his diary for 2 February 1840 a remarkable evening spent with Wheatstone:

Last night I was hardly able to sleep, from the strong impressions made on my mind by the stupendous discoveries and results of experiments by Mr. Whetstone [sic] on electricity, and his most ingenious mechanical apparatus for an electric telegraph. He had kindly met me by appointment in the lecture room of King's College, and for an hour and a quarter was incessantly occupied in explaining to me alone the whole doctrine, and the admirable application of it to this purpose of a telegraph. The velocity with which the communication takes place is almost inconceivable. By some curious experiments, however, he seems to have ascertained that it travels 160,000 miles, or more than eight times the circumference of the globe, in one second; and what is more wonderful still, he speaks of this, not as denoting instantaneous or immeasurable, but he has contrived to measure a subdivision of time equal to one-millionth part of a second, and he speaks of this one second as a portion of time ascertained, so that it might take two seconds to travel

400,000 miles, and so on. Gas and steam have done much, but this agent is destined to do much more, and to work an incalculable change in human affairs. It far exceeds even the feats of pretended magic, and the wildest fictions of the East.[8]

The figures quoted in the diary (160,000 miles in one second and 400,000 miles in two seconds) are not self-consistent, but they suggest that Wheatstone was then prepared to give an estimate of the velocity and that his estimate was less than the 288,000 miles per second mentioned in his 1834 paper.

On 6 February 1840, a few days after his conversation with Copleston, Wheatstone gave evidence to the House of Commons Select Committee appointed to 'Inquire into the State of Communication by Railways'.[9] The Committee decided that they needed to consider the newly developing telegraph as well, and they called Wheatstone to give evidence. In answer to a question about the time needed to send a telegraph message, Wheatstone said 'From some experiments I made some years ago . . . I ascertained that electricity travelled through a copper wire at the rate of about 200,000 miles in a second'. He later gave the speed of light as 192,000 miles in a second.

In October 1840 Wheatstone's Belgian friend, L.A.J. Quételet, writing about Wheatstone's telegraph in the *Bulletin of the Royal Academy of Brussels*, referred to electric signals as travelling around the earth six or seven times in a second, which implied that Wheatstone was then speaking of the velocity of electricity as about 150,000 to 180,000 miles per second – a little less than the speed of light. When Wheatstone's paper on the measurement of the velocity of electricity was published in a French translation in 1842, the figures he had given in 1834 were not altered, and no later results were added.

An obvious question to ask is whether the results Wheatstone obtained were correct. His measurement of the time interval between the sparks was probably accurate, but he was not really measuring the flow of current along the whole length of wire. Wheatstone could not have understood at that time the effects of capacitance between the wires, and the spark system he was using would have led to a very rapid rise and fall of the current in his wire. What he was actually measuring was the speed with which an electric signal was transmitted through an arrangement of capacitors. But he also found what he needed to know: electricity travels very fast, and he could operate his instruments through very long lengths of wire.

The speed of light

Wheatstone considered the possibility of measuring the velocity of light through the atmosphere by means of a revolving mirror.[10] Since the velocity of light was known at the time only from astronomical observations, it would have been of great scientific interest to make a terrestrial measurement.

The proposed method was to observe two images of a spark in his revolving mirror. The first image was obtained by a single reflection of the spark in the revolving mirror; the second was obtained from the first by reflecting the first image on to a fixed mirror and then back to the revolving mirror. The spark was to be generated at the instant the two mirrors were parallel, but because the revolving mirror would have turned slightly before the light reached it for the second time, the second image would be displaced from the first. There is no evidence that Wheatstone actually tried out this experiment – indeed it is most unlikely that he did, since a little arithmetic would have shown him that it would be quite impossible with his apparatus. The velocity of light is nearly three hundred million metres per second. Copleston's diary quoted above implied that Wheatstone claimed to be able to measure a period of one millionth of a second. Even if he could do that with any accuracy, he would have required the fixed and revolving mirrors to be at least 150 metres apart, and King's College has no corridor of that length. The light of the spark would have had to travel at least 300 metres; it is improbable that Wheatstone could have generated a precisely timed spark which was bright enough to be seen at that distance, and inconceivable that the whole experiment could have been conducted out of doors on a very dark night.

Wheatstone's proposal, and a variant of it, are preserved in a manuscript note among his papers at King's College:

To measure the Velocity of Light
1st Mode

Place before the rotating mirror a charged jar and let it be discharged at the moment the mirror is perpendicular; at a considerable distance before it place a vertical, stationary, plane mirror. In the rotating mirror will be seen the reflection of the spark, and also the second reflection from the stationary mirror; the angular distance of the two sparks when the mirror is stationary, and when it is in motion, will indicate by their difference the time in which light travels through double the distance from the spark to the mirror. One of the great difficulties in this apparatus will be to discharge the jar at the precise instant the mirror is vertical.

If the stationary [mirror] be thus placed AB [see Figure 6.6] the eye placed before the rotatory mirror when stationary will see one image superpose [sic] the eye, and the degree of separation will indicate the velocity. The spark being placed at S and the eye at ⊙, let a line ⊙S' be drawn to the image of S; that the two reflections may superpose the mirror AB must be parallel to this line.

Instead of a reflecting surface, two sparks at the extremities of a long wire placed vertical to the mirror may be taken; in this case the time of electricity passing must be deducted.

2nd Method

The method by the revolving mirror seems favourable from the facility of multiplying the apparent velocity by taking successive reflections in two revolving mirrors; but the difficulty of estimating the angular deviation induced me to look for another method.

For either of the preceding experiments substitute for the revolving mirror a small circular disk with a white surface and a fixed black radius. If there is any difference of time between the two sparks, in the one case, or the spark and its reflected image in the other, when the disk is in motion two radii will be seen and their angular distance will indicate the time. The sensibility is not so great as the mirror because the angular deviation cannot be observed so accurately, and in one revolution of the disk only one such is made. The following arrangements may however in some measure overcome this imperfection.

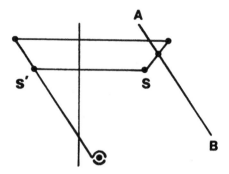

Figure 6.6   Wheatstone's sketch from his notes about measuring the velocity of light.

Instead of a disk cause to revolve only a single radius, but instead of regarding the radius itself look at its lengthened shadow cast on a card placed at right angles to the plane of its motion; or observe the radius itself by a telescope with a micrometer. Or make the revolving radius an object in a solar microscope, but illuminated only by the electric light.

This second method thus conceived by Wheatstone was employed by Foucault, some 15 years later, when he used a rotating mirror, and a much brighter light-source. He made the first terrestrial measurement of the velocity of light, but acknowledged that the method was Wheatstone's.

Notes

1   Alfred J. Frost, *Biographical Memoir of Sir Francis Ronalds, FRS*, included in *The catalogue of books and papers relating to electricity, magnetism, the electric telegraph, etc. including The Ronalds' Library*, 1880.
2   Letter from Wheatstone in Royal Society Archives
3   The article by Fox Talbot is in *The Philosophical Magazine*, 3 August 1833, pp. 81–2. Wheatstone's letter appeared in the September issue, pp. 204–5.
4   E.T. Whittaker, *History of the Theories of Aether and Electricity*, first published in 1910. The quotation is from the 1951 edition, p. 227.
5   See the appendix with Wheatstone's publications for full bibliographical details.
6   King's College Council Minutes, 30 June 1834.
7   Correspondence with Roget in Royal Society Archives
8   W.J. Copleston, *Memoir of Edward Copleston DD, Bishop of Llandaff*, 1851, p. 169.
9   *Minutes of Evidence taken before the Select Committee on Railways*, House of Commons Sessional Papers 1840 (xiii). Minutes 334 and 336.
10   Report by Quételet in *Bulletin de l'Académie Royale des Sciences et Belles-lettres de Bruxelles*, **7**, 1840, pp. 131–4.

Chapter 7

Professor Wheatstone

Kings College London was one of a number of educational establishments founded in the first half of the nineteenth century to meet the increasing demand for a wider education than that offered by the universities of Oxford and Cambridge. Those universities imposed strict religious tests on their students and provided only a conventional education with no science.[1]

As industry developed there was an ever increasing need for skilled working men with an understanding of the rudiments of science and technology. The Royal Institution had been founded in 1799 for diffusing and facilitating the general introduction of useful mechanical inventions and improvements and for teaching by courses of philosophical lectures and experiments the application of science to the common purposes of life. The London Mechanics Institution, which developed into Birkbeck College, was established in 1823 by Dr George Birkbeck (one of the people to whom Örsted had shown Wheatstone's early work). Many similar Mechanics Institutes were formed around the country, and their success emphasized the need for some provision for middle-class higher education and led to several proposals for new places of learning which could offer an education comparable with that given by the old universities, but covering a wider range of subjects.

The first University of London opened in October 1828. It was founded by an amalgamation of two distinct groups of people. One was a group of Non-conformists led by the Presbyterian Edward Irving and the Baptist Dr F.A. Cox; the other was a radical and mostly atheist group including Jeremy Bentham and John Stuart Mill. The new university could not confer degrees because it had been unable to get a charter. The students (300 in the first year) were offered an education in such practical subjects as medicine, law, mathematics, natural science and political economy, as well as the classics.

Before this 'University of London' had opened there were moves among leading Anglicans to establish a college whose syllabus, whatever else was taught, would include Christianity according to the doctrines of the Church of England. In June 1828 a prospectus was issued for a college, to be known as

King's College. The inaugural meeting on 21 June 1828 was presided over by the Prime Minister, the Duke of Wellington, accompanied by the Archbishops of Canterbury, York and Armagh, and seven bishops. The new college was initially a private company (as was the university), and within two months the money raised or promised exceeded the initial target of £100,000. The site, adjoining Somerset House in the Strand, was not determined for another year; other suggestions which were considered included Buckingham Palace, not then the Sovereign's residence, and a site in Regent's Park later occupied by Bedford College and now by Regent's College. Building began in 1829 and the college opened on 8 October 1831. King's College was frequently in financial difficulty, partly because many ultra-Protestant supporters withdrew their support in protest at the Catholic Emancipation Act, passed by Wellington's Government in 1829.

King's College was granted a charter in 1829, converting it from a private company to a public institution, but still not conferring the right to grant degrees. The desire for a body in London with power to grant degrees led to the initial establishment of the present University of London in 1836 as a purely examining body. At the same time the first University of London received a charter also and changed its name to University College London.

King's College was originally intended to have 28 rooms for the professors, and 10 lecture rooms with a capacity of 2,000 students. It opened with 6 lecture rooms, 18 for professors, a chapel to seat 800 and basement schoolrooms for 400 boys.

The College was divided into the Lower Department (now King's College School at Wimbledon) which provided an education for boys up to the age of 16, and the Upper Department whose function was to prepare young men for commercial and professional life. King's also prepared students for admission to Oxford and Cambridge. The Upper Department was much more like a modern Sixth Form than what is now understood by a university college. The control of the College was entirely vested in its Council; the principal and the professors had no part in it. The staff, except the Professors of European Languages, had to be members of the Church of England. No religious test was applied to the students but they had to attend services in the College chapel. Most of the professors received no fixed salary, but took a proportion of their students' fees.

In 1834 Wheatstone was appointed Professor of Experimental Philosophy when the Reverend Professor Henry Moseley, who had held the chair of both Natural and Experimental Philosophy (that is, theoretical and practical physics), gave up the experimental side and became Professor of Natural Philosophy and Astronomy. There is no record of how Wheatstone began his connection with King's College, but before he joined the staff he was already known to at least one of the professors, Herbert Mayo, the first Professor of Anatomy. Mayo had published a revised edition of his *Outlines of Human Physiology* in 1833, the year before Wheatstone's appointment, and in this he devoted a page to Wheatstone's work on vision and his explanation of what was subsequently called stereoscopic vision. Wheatstone did not publish anything on this subject until

Figure 7.1   Wheatstone in 1837. Drawn in chalk by William Brockedon.

Source: By courtesy of the National Portrait Gallery, London.

1838, and later he said that his work on stereoscopy was first made public by Mayo.

Wheatstone was so shy by nature that it is difficult to imagine him applying for a teaching post. It seems in fact that he did not apply – his services were sought. In a letter to Sir John Herschel, who was in South Africa to make astronomical observations in the southern hemisphere, Wheatstone wrote on 13 May 1835:

Since your departure I have been appointed Professor of Experimental Philosophy in Kings College, London; in this appointment, which was quite unsolicited on my part, your friend Prof. Jones took considerable interest.[2]

Professor Jones was Richard Jones, Professor of Political Economy at the College. For some professorial appointments, the post was advertised and a selection committee set up, but there is no mention of such procedures in the College Council minutes at Wheatstone's appointment. The minute reads:

The Principal having communicated Prof. Moseley's consent that a Professor of Experimental Philosophy should be appointed, he retaining the Professorship of Natural Philosophy and Astronomy, *Resolved* that Mr. Charles Wheatstone be accordingly appointed Professor of Experimental Philosophy.

At first sight it may seem surprising that the man from the music shop down the road was appointed to the College post, but in fact he had been well educated compared with many of his age and class. Faraday, for example, learned little more than the rudiments of reading, writing and arithmetic at a common day school. His science was largely self taught in such free time as he could make during his bookbinding apprenticeship. Wheatstone's schooling had included mathematics, physics and languages; the books his father obtained for him and his experiments in the scullery gave him a good scientific education. He clearly had his father's backing for his studies at an age when he might reasonably have been expected to earn his own living. His achievements had already justified his father's confidence in him, proved profitable to the family business, and won him a place in the scientific community.

The precise duties of the professors were somewhat vague. It was certainly not a full-time job, and Wheatstone continued with his musical business. When he obtained a patent in 1836 for musical instruments he still described himself as Charles Wheatstone, Musical Instrument Manufacturer, of 20 Conduit Street. (In subsequent patents this became first Charles Wheatstone, Esquire, then Charles Wheatstone, gentleman.) The Professor of Geology, John Phillips, lived at York, which was apparently no bar to holding the post, and he employed a lecturer to take classes for him. The Professor of Chemistry, J.F. Daniell, combined this office with being lecturer in chemistry and geology in the East India Company's Military Seminary at Addiscombe, and he was also foreign secretary of the Royal Society. Some mundane administrative work fell upon the professors, and among Wheatstone's papers is a letter dated May 1839 from the College Secretary, conveying a message from Sir Robert Smirke, architect of the College, about new hat rails for the students.

Six weeks after his appointment, the Council granted him £50 for providing apparatus for the use of his lectures. He spent this during the next few months and subsequently told Herschel in the letter referred to above:

At present I am very badly provided with apparatus; but still I have some additional facilities to prosecute experimental researches, and this is an advantage to me, though as no pecuniary profit at present results I am at present unable to devote so much time to the duties as I would wish to do.

There was no further grant of money by the Council. In December 1835 they turned down a proposal that they should fit up a room for receiving Professor Wheatstone's apparatus. The nature of the apparatus is unspecified, but in the following February he was granted permission to lay wires in the vaults for his velocity of electricity measurements at the expense of the Royal Society. There is no further mention of Wheatstone in the College Council's minutes until after his death, 40 years later.

Wheatstone retained the Chair for the rest of his life. It is far from clear what he actually did at King's, but after his death the College Council held him in such high esteem that they resolved that the physical laboratory of the College should be distinguished by the name of the Wheatstone Laboratory. The College has never honoured anyone else in this way. There is no complete record of the lectures given by Wheatstone (or any of the professors) at the College, but two printed lecture syllabuses survive among his papers in the College library. One is for a course of eight lectures on sound, given at an unspecified time on successive Tuesdays, beginning 17 February 1835. The other is also for a course of eight lectures at eight o'clock in the evenings of successive Tuesdays, beginning 9 May 1837, on the subject of 'The Measures of Sound, Light, Heat, Magnetism and Electricity'. Presumably these lectures were attended by part-time students after a day's work.

The anonymous writer of Wheatstone's obituary in the *Proceedings of the Royal Society* wrote that after his appointment 'he delivered a course of eight lectures on Sound in the early part of the following year; but his habitual though unreasonable distrust of his own powers of utterance proved to be an invincible obstacle, and he soon afterwards discontinued his lectures'. The eight lectures on sound seem to have been the sum total of his lecturing during his first year at the College. Presumably his contact with students was mainly in tutorial classes and in the laboratory. Whether or not the sound lectures were repeated at the College in 1835, Wheatstone certainly gave a series of lectures on sound at the Royal Institution in 1836. The programme card gives no details of the syllabus, but includes in the list of lectures to be delivered after Easter:

Sound – by Charles Wheatstone, Esq., Professor of Experimental Philosophy in King's College London. To commence on Tuesday, the 3rd of May, at Three o'clock, and to be continued on each succeeding Tuesday till the 7th of June.

The substance of the lectures was presumably the same material as delivered at the college, although now condensed into six parts instead of eight. Wheatstone did give the lectures at the Royal Institution, for they are included in a list of lectures actually given in the managers' minutes. The other courses of lectures on the same programme card were on heat, by Faraday; on the physiology of the senses, by Roget; on botany, by Lindley; on early English opera, by Edward Taylor; and on landscape painting, by Constable. The Royal Institution records show that three non-members paid one guinea each to attend Wheatstone's course. There is no record of the number of members who attended. Faraday attracted 60 non-members, Lindley, Taylor and Constable four each, and Roget none.

The Historian of King's College, Professor F.J.C. Hearnshaw, made two observations on Wheatstone in his *Centenary History of King's College London*, published in 1929. Hearnshaw's assessment of Wheatstone's place in the College history is a valuable judgement, based on a thorough study of that history:

Professor Wheatstone was making a great sensation in the world of science and technology by his amazing discoveries and inventions in electricity. He had wholly ceased to give lectures to students and was entirely engrossed in fruitful research. He was the first of a notable series of great men associated with King's College who gradually converted the institution from a mere higher grade school, concerned only to purvey existing knowledge, into a true university wherein not only is the known imparted, but the unknown explored . . .

In 1875 the great Sir Charles Wheatstone died. For 41 years he had been nominally professor of experimental physics in the college. After his first session, however, he had done no lecturing, but had devoted himself (presumably without salary) to his researches in electricity, and to his inventions in telegraphy, magneto-electrical machines, and recording instruments. On his death he left a Will in which he bequeathed to the corporation of King's College all his scientific books and apparatus, all his medals and diplomas, together with the sum of £500 to be expended in laboratory equipment. The books numbered 1500; the apparatus, which included much of that with which Sir Charles had made his pioneer experiments, was estimated to be worth at least £1000. The council gratefully accepted the generous bequest, decided to erect a special gallery in the George III museum to receive and display the apparatus, and, having spent the monetary portion of the legacy on scientific equipment, named the physical laboratory in which it was placed the 'Wheatstone Laboratory'.

The gift of £500 for laboratory equipment may be contrasted with the £50 he received for apparatus when he first went to King's. Clearly Wheatstone thought that the provision of adequate laboratory apparatus was important. He also left the Royal Society £500 to be applied to the fund from which he had received £50 for the velocity of electricity experiments. The Wheatstone Laboratory still bears his name and his apparatus still belongs to the College, although much of it is on loan to the Science Museum. His books remain with the College.

Hearnshaw is not quite correct in stating that Wheatstone had done no lecturing after the first session, though Wheatstone himself may have encouraged the idea that he could not lecture. In 1859 Sir James Kay-Shuttleworth invited him to speak in public about his latest telegraphs. Wheatstone replied: 'I never addressed a public meeting in my life, and never can.' That was an exaggeration but it reveals his attitude to speaking in public. The legend of Wheatstone running away from the Royal Institution may have coloured the recollections of several people who remarked upon the contrast between Wheatstone in private conversation and Wheatstone speaking in public. The friend who wrote the obituary memoir for the Royal Society said:

anyone would be charmed by his able and lucid exposition of any scientific fact or principle – yet his attempt to repeat the same process *in public* invariably proved unsatisfactory.

W.H. Preece said:

He was no lecturer himself, yet no one was more voluble in conversation. At explaining any object of his own invention, or any apparatus before him, no one was more apt, but when he appeared before an audience and became the focus of a thousand eyes, all his volubility fled . . .

His appointment was not universally popular. Leopold Martin, son of the painter John Martin, who was a friend of Wheatstone's, published *Reminiscences* about his father in which there are several references to Wheatstone. He referred to his appointment at King's as:

an appointment which was hardly popular, one may say, at the College, as he failed as a lecturer. The matter or treatment might be clever, but his utterance was so rapid that it was nearly impossible to follow him; hence the want of popularity with his class.[3]

Probably there was no obligation on the professors to give formal lectures. Wheatstone would have been an excellent tutor to individual students. He retained the college post for the rest of his life, and was one of the people chiefly responsible for developing the tradition of research which helped to change the King's College from a senior school to an institution of university status.

In October 1839 King's College initiated a course in civil engineering.[4] The Council had noted the lack of educational facilities for people wishing to become civil engineers, and in 1838 they decided to introduce a suitable course of instruction, having special reference to the arts and manufactures of the country. At that time the term civil engineering included all non-military engineering. The course lasted three years and the fees were £31 10s. per annum. Each professor gave one lecture per week for one or two terms in each year of the course. Individual students could attend any one professor's course for £1 1s. per term. The civil engineering students were quite distinct from the regular students of the College and the engineering lectures were in the evenings. Examinations led successful students to the associateship of King's College. The syllabus comprised:

Mathematics, Mechanics, Hydrostatics, and Hydrostatical Machines; the Steam Engine and its applications; theoretical and practical chemistry; Metallurgy, Geology and Mineralogy; the theory and practice of mining; the elementary properties of Matter, Sound, Light, Heat and Electricity; Machine-drawing, practical Perspective and Surveying.

These were all subjects which could be taught by the existing staff and it can be

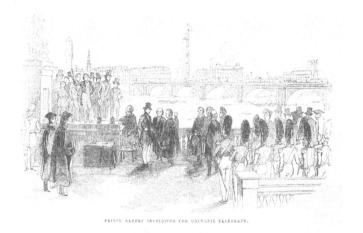

PRINCE ALBERT INSPECTING THE GALVANIC TELEGRAPH.

Figure 7.2   Prince Albert inspecting the telegraph at King's College in 1843.

Source: *Illustrated London News* of 1 July 1843.

assumed that Wheatstone was responsible for the sound, light, heat and electricity. It was intended that:

The observation, judgement and invention of the students will be exercised by experiments made by themselves, and by visits to various manufactories and other works, to which access has been liberally granted . . . and where they will be accompanied by the lecturer, who will give explanations on the spot.

A study of the annual College calendars shows that the engineering course continued throughout Wheatstone's lifetime. His influence may be seen in the syllabus, and since no one else among the staff would have had comparable knowledge of the electrical subjects he must have been involved in the teaching, presumably by practical classes and informal groups. No doubt Wheatstone 'liberally granted' access to his own premises in Conduit Street and showed his students the manufacture of musical instruments as well as his acoustic and electrical experiments. Away from the lecture room and with a small informal group of students his powers of able and lucid exposition would have been of the greatest benefit to the pupils.

The *College Calendar* for 1864–1865 may be quoted as an example of the calendars from about 1850 to 1875. The course in the Department of Applied Sciences – Engineering Section includes under the heading *Electricity* the following topics: The Electric Current; How to measure it; The Resistance it meets with; The Force which sustains it; Construction of Telegraph Lines. The heading *Manufacturing Art and Machinery* includes the subheadings of *Ropemaking*, which includes Hemp, Heckling, Spinning, Forming, Laying, Wire Ropes, Telegraph Cables, and *Telegraphs* Semaphore, Electric, Needle, Dial, Post Insulators, Submerging Telegraph Cables. No one among the College staff but Wheatstone could have dealt with the practical aspects of electric telegraphy covered in this syllabus.

Wheatstone's place in the College after his first few sessions was probably similar to that of a modern research fellow. In the calendars he is duly listed among the professors at the beginning of each calendar, but his name does not appear with the teaching staff of any department.

About this time the College embarked on publishing a series of textbooks under the title *Illustrations of Science by Professors of Kings College*. These were intended to appear at intervals of three months and were to form a complete course of instruction in natural philosophy and natural history. The first of the series, Professor Moseley's *Illustrations of Mechanics*, appeared in March 1839 and was also apparently the last. According to advertisements of the time the first volume of Professor Bell's *Illustrations of Zoology* was due in June, and Volume I of the *Illustrations of Experimental Philosophy* in two volumes by Professor Wheatstone was due in September 1839, but neither book was ever published. However, some of the College staff did publish textbooks and these sometimes bore the College coat of arms on the title page. An example is Daniell's *Introduction to Chemical Philosophy*, published in 1843.

There are several references to Wheatstone in the reminiscences of Faraday published in 1931 by M.C. Grabham (1840–1935), a very old former student of King's College. Grabham was the youngest of four brothers who all went to the Lower Department of the College, now King's College School. Grabham wrote:

Professor Wheatstone made practical application of Faraday's experimental results as fast as they were issued . . .

Faraday was constantly in and out at King's College in my early years of 1849, '50 and '51. He would have long talks with the Chemist, Prof. Daniels [sic], who had just then brought out his copper sulphate *constant* cell – a dictation rather than a talk; but his visits were chiefly concerned with Prof. Wheatstone, mostly brief, with a few emphatic words, simple, but almost dictating . . .

Wheatstone and Daniels, both highly cultured men, always seem in my memory to have been suppressed and silent in Faraday's presence; he was emphatic, never dogmatic; and little reply was ever made to his assertion of facts or suggestion.[5]

Grabham's memory was fallible after 80 years, for Daniell died in 1846. Perhaps he was combining his own memories with what he had been told by his elder brothers and fellow students. But his account carries conviction and, by some means he does not explain, he acquired some of Wheatstone's apparatus after his death. These items were subsequently presented to the Science Museum. Grabham also gave an interview to a correspondent from *The Times*, in which he said that, after Faraday had been to see him, Wheatstone would sink into a reverie in which he seemed unaware of anything around him.

After Wheatstone's death a few box-files of his papers were retained in the College library. These are mostly odd notes and only occasionally of interest. One group of papers that does merit special attention is the manuscript – or at least most of it – of his course of lectures on sound. The manuscript contains many alterations, and it is not always possible to determine which passages are rejected drafts and which represent the lecture in its preferred form. Lecture I, most of whose papers are clipped together, consists of about 6,000 words on 31 quarto pages. Lecture III is headed also 'A lecture on the vibrations of columns of air in cylindrical and conical tubes delivered at the Royal Institution March 15th, 1832.' Presumably that earlier lecture was incorporated into the course. The opening sentence of Lecture V indicates that it lasted one hour.

The manuscript is interesting because it reveals the depth of coverage of the subject. The first lecture discusses the causes of sound, the factors determining pitch and timbre, hearing, the determination of frequency, the mathematical basis of harmony and the equally tempered scale (but he does not mention his Harmonic Diagram). He quotes Galileo, Hooke, Young, Cagniard de la Tour, Chladni and Savart, among others. Clearly Wheatstone was entitled to claim, as he did in the letter to Sir John Herschel of 13 May 1835, that he had 'concluded a course on Sound, in which I endeavoured to include every experimental investigation which has hitherto been made in this department of physics'.

Notes

1 The main source for the early years of King's College is Professor F.J.C. Hernshaw, *Centenary History of King's College, London, 1828–1928*, 1929. The author was Professor of History at the College.
2 Royal Society Archives, Herschel correspondence.
3 Leopold Martin's 'Reminiscences' were serialized in the *Newcastle Weekly Chronicle*. The extracts quoted are from the issues dated 30 March and 20 April 1889.
4 The new Civil Engineering Class at King's was advertised and reported in journals of the time including the *Mechanics' Magazine*, vol 32, 26 October 1839.
5 M.C. Grabham's *Recollections of Faraday* were printed privately on the occasion of the British Association meetings during the Faraday Centenary Celebrations in 1931. His interview with *The Times* appeared on 21 September 1931.

Chapter 8

Researches in electricity

When an electrical engineer today speaks of generating electricity, his mind conjures up a picture of rotating machinery with coils and magnets. The idea that most electricity comes from such generators is so ingrained that it is difficult to appreciate that before the appearance of the self-excited dynamo in 1867 it was by no means certain that magneto-electric induction was destined to provide most of the world's electric current.[1]

Edward Highton, in his *History and Progress of the Electric Telegraph*, written in 1852, had a section 'On the Production of Electricity', in which he said:

Electricity may be produced in a variety of ways: by friction; by chemical action; by magnetic induction; by change of temperature; by the power and at the will of certain animals.

He said that although those were the principal sources from which electricity in large quantities could be obtained, they were by no means the only ones. He devoted much the same space to a consideration of each of these sources, and concluded that although only chemical batteries and magneto-electricity had so far been successfully applied to the telegraph, the use of thermo-electricity was probable. He also recommended that telegraph engineers should study the electric fish, for if they could understand the working of the fish when wholly submerged in water, they would probably discover 'a means of constructing submarine telegraphs, without any insulation of the wires'. Such an invention would indeed have been valuable!

It is against this background of several different possible sources that we must view the work leading up to a practical means of generating large quantities of electricity. Wheatstone took an interest in several of these possible sources. In 1837 he published a paper in the *Philosophical Magazine* under the title 'On the Thermo-electric Spark'. It illustrates the value of his knowledge of other languages and his wide reading of foreign scientific journals. An Italian journal had given an account of experiments with thermo-electricity by the Florentine scientist Antinori and by Professor Linari of Sienna.[2] Wheatstone repeated what he considered to be the most significant part of the experiments and in his paper he gave an account of both the Italian report and his own confirmatory work. Antinori had succeeded in obtaining various electric effects from a thermo-electric pile, including an electric spark obtained by interrupting the current in an inductive circuit. This was significant because it was by no means certain at that time that the 'electricities' obtained from different sources were identical. Faraday had shown in 1832 that electricity from a voltaic cell and the electric charge on a Leyden jar were of the same nature and produced similar chemical effects. Antinori reported producing magnetic and chemical effects as well as the electric spark with electricity from a thermopile. Wheatstone remarked that Antinori had provided 'a link which was wanting in the chain of experimental evidence which tends to prove that electricity from sources, however varied, is similar in its nature and effects'.

Wheatstone had two thermo-electric piles made by John Newman, the instrument maker. One consisted of a cylindrical bundle of 33 bismuth and antimony elements about three centimetres long. One end was cooled with ice and the other heated by a nearby red hot iron. The pile was connected in series with 15 metres of copper ribbon insulated with brown paper and silk and wound into a spiral. The circuit was completed by wires dipping into a vessel of mercury, and on withdrawing one of the wires from the mercury a spark was visible – even in daylight – as the circuit was broken. Wheatstone's friend Daniell and the American Professors Henry and Bache, who were visiting Wheatstone in the course of a European tour, took part in the experiment, which Wheatstone later showed to Faraday. Henry gave a lively account of his visit to Wheatstone in his *Journal of European Trip*, 1837. He says that when he went to London he resolved to repeat the attempt to obtain a spark from thermoelectricity:

Mr. Wheatstone, however, informed me a few evenings since that he had seen in an Italian journal an account that the spark had been obtained from thermoelectricity by means of a spiral. Dr. Daniell thought it would be impossible to obtain it with the means at command. I, however, prepared the apparatus and made the attempt by holding a hot poker to one side of the pile and ice to the other. The pile, the coil and a small movable piece of wire formed the circuit which could be broken by drawing out the end of the movable wire from the mercury cup. The heat and cold was managed by Prof. Wheat-stone, the parts of the apparatus were held by Prof. Daniell. I managed the wires. All ready – no effect. Again – no effect. And again the same result. The other gentlemen now withdrew to another room to inspect some letters. I made a new arrangement of the apparatus, adjusted every part more carefully, called the others again, got all prepared. Prof. Wheatstone as before applying the ice and Prof. Daniell the poker. 1st attempt, no effect perceived; the second, a small spark each time I broke the connection. This was seen by all in succession. Bache had arrived just before our first attempt and had gone into another room to inspect the letters above alluded to. He requested to be called if the experiment succeeded. The noise we made on the occasion called him forth. He afterwards made much sport with our enthusiasm: Prof. Daniell flourishing the poker, Wheatstone with the ice, and I jumping as he said in extasy [sic]. The experiment is, however, an interesting one and not the less so that we are the first mortals who have witnessed it in England.[3]

There Wheatstone appears to have left the subject of thermoelectricity after expressing the view that, with suitable combinations of metals, it might be possible to produce a thermoelectric pile having an effect equal to that of an 'ordinary voltaic pile'. But if Wheatstone had been asked at that time to predict which sources of electricity would be important in the future, he would probably have said the choice lay between the chemical battery and the thermopile. Edward Highton, in the book quoted above, made the prophecy that 'the day is not very distant when a farthing rushlight will be capable of developing sufficient electricity to keep up an instantaneous communication by telegraph between London and Liverpool'. Today the direct generation of electricity from heat, without the intervention of a boiler producing steam to drive a turbine and generator, is still a dream of electric power engineers. Thermoelectric generators are used for some low power applications, such as in satellites where cost is not important. In 1838 even the most far-sighted person could not have foreseen the quantity of electric power which would eventually be available, and thermoelectricity must have seemed as promising a source of power as any other. Wheatstone studied this as well as magneto-electric devices and electric fish.

Faraday's discoveries in electromagnetic induction were made in 1831, but their overriding significance was only realized much later. It is with hindsight that Faraday is seen as 'the father of electricity'. Faraday himself conducted a series of experiments with electric fish at the Adelaide Gallery towards the end of 1838 and Wheatstone was present at several of these. After giving an account of his own experiments on one such occasion Faraday continued: 'Tried Mr. Wheatstone's apparatus. It did not do. It is beautiful in principle but requires more electricity'. Unfortunately there is no record of what 'It' was or what 'It' did not do! But from the context of Faraday's remark presumably it had something to do with the electric fish.[4]

Cooke and Wheatstone experimented with magneto-electric devices while preparing their first joint patent specification for an electric telegraph, which was filed in December 1837. Cooke wrote 'We mean to try whether a machine called "Electro-Magnetic" cannot be made to supersede the Galvanic Battery'. Since their patent specification is silent on the subject, the answer must have been 'no': at that stage they were not able to work their telegraph with a magneto-electric device.[5]

In 1836 Wheatstone's friend and colleague J.F. Daniell (1790–1845), Professor of Chemistry at King's College, devised the primary cell known by his name. The Daniell cell, which did not polarize and so gave a more constant current than previous types, consisted of zinc placed in zinc sulphate solution and copper placed in copper sulphate solution, the two solutions being separated by a porous earthenware partition.

Wheatstone also devised a 'constant' cell, using a single electrolyte. This was a variant of Daniell's cell and he described it in the paper on electrical measurements which brought 'Wheatstone's Bridge' into prominence. This cell had a porous pot of half-baked clay filled with a paste of zinc amalgam and standing in a glass or porcelain jar of copper sulphate solution. The negative pole was a copper wire in the amalgam and the positive pole a copper plate in the copper sulphate.

The cell was used in practice for some time after Wheatstone's death. Its use is advocated in resistance measurements by the bridge method in *Practical Electrical Engineering*, a two-volume tome claiming to be 'A complete treatise on the Construction and Management of Electrical Apparatus as used in Electric Lighting and the Electric Transmission of Power', which was written in about 1890 when public electricity supply was less than ten years old.

Generators

In January 1840 Cooke and Wheatstone obtained a patent for an electric telegraph which could use either a battery or a magneto-electric machine for its source of power. Presumably they had by then a magneto which was more satisfactory than those they had used in 1837. According to their patent specification, which was written by Wheatstone alone, the battery-driven version was preferred and the magneto version was an alternative, but within a few years the magneto version became the favourite.

As well as this telegraph instrument there is still in existence a hand-driven magneto of Wheatstone's which is so similar that it must have been made at the same time. This machine is in working order. If the handle is turned at one revolution per second – which is hard work – the current generated in a short circuit is about five milliamps. The open circuit voltage is about 16. There is a commutator and the output is a series of unidirectional pulses. The commutating arrangements seem to have given Wheatstone a great deal of trouble in practice, and the commutators of these machines have to be kept spotlessly clean. Many of his later magnetos were designed to give alternating outputs so that commutators would not be required.

To set Wheatstone's work in context it is helpful to look at the development of the magneto before 1840. Faraday made the basic discovery of magneto-electric induction in October 1831 when he showed that an electro-motive force is induced in a coil of wire as a magnet moves towards or away from it. The following month he made his disc generator, with a copper disc rotating between the poles of a permanent magnet. The copy of it in the Science Museum, which uses the magnet actually employed by Faraday, generates a current of about seven micro-amps. The magnet has probably deteriorated since Faraday's day, but this machine was not destined to replace the chemical battery.

Generators working on the same principle as Faraday's disc generator ('homopolar' machines) are sometimes used where very heavy currents and low electro-motive forces are required. Except for that very specialized application, however, all generators have magnets and coils of wire which rotate relative to one another. Either permanent magnets or electromagnets may be used. The first generator of this kind was the machine made by the Paris instrument maker Hippolyte Pixii in 1832, which had a rotating permanent magnet and fixed coils.

At first there was no commutator and the output was alternating, but a pivoted switch was soon added, to act as a commutator. Pixii's machine was exhibited at the Adelaide Gallery in London in November 1833. (Only two complete Pixii machines are still in existence: one in the Smithsonian Institution in Washington and one in the Deutsches Museum in Munich. The Science Museum has a copy of the former.)

In Saxton's machine of 1833 the coils rotate and the magnet is fixed, but the relative positions of coil and magnet are unchanged. The commutator is a metal strip which dips into a cup of mercury. In Clarke's machine of 1835 the coils rotate alongside the magnet rather than at the end of it and the commutator resembles that on a modern machine. This was the arrangement adopted by Wheatstone in the magneto of his 1840 ABC telegraph. There was nothing novel about Wheatstone's magneto, but the way it was used for a telegraph was quite new.

After the discovery of magneto-electric induction a number of reciprocating generators were made, in which a magnet was moved rapidly in and out of a coil by turning a handle. There is one such machine among the Wheatstone relics at King's College: it has two coils which move between the poles of two compound horseshoe magnets placed end to end. There is no mention of this machine in any of Wheatstone's publications.

All the machines described so far had an output waveform consisting of a series of unidirectional pulses. The first object of the patent Wheatstone obtained in July 1841 was to make a magneto-electric machine whose output is 'not to be distinguished from a perfectly continuous current'. He thought that such a machine could be used for many purposes for which a voltaic battery was employed, especially to produce those effects which required a battery consisting of a considerable number of small elements.

The new machine, which Wheatstone called a 'magneto-electric battery' (Figure 8.1), had five armatures rotating on a common shaft between six

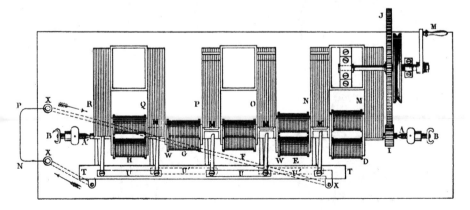

Figure 8.1   *Wheatstone's 'magneto-electric battery', which had five sets of armatures, each with a commutator, arranged at 72° intervals on the shaft.*

compound horseshoe magnets. Each armature had its own commutator and the armatures were each displaced relative to the next by one fifth of a revolution. The outputs from the five armatures and commutators were connected in series, and the commutators were made so that the circuit was never broken as the wiper passed from one commutator segment to the next.

An obvious question to ask is why Wheatstone adopted five armatures, rather than four or six? The specification does say that more or fewer than five armatures could be employed, but five was definitely the preferred number. It is possible that Wheatstone experimented with varying numbers of simple machines linked together. Using a lightly damped galvanometer to observe the output, he could easily have determined which arrangement gave the smoothest output. I have observed the output waveform of Wheatstone's hand-driven 'ABC Telegraph' magneto with an oscilloscope. The resultant of five such waveforms, equally spaced, is fairly smooth (Figure 8.2). No machine of this design is now extant, and it is impossible to tell how many were made but there seem to have been more than a single prototype. When discussing early magnetos in 1873 Du Moncel referred to it as 'the Wheatstone machine', without mentioning any other by Wheatstone.[6] This suggests that it came as near to being a practical machine as any magneto of 1840 could.

There were other multi-polar machines made at this time – such as Stöhrer's machine – but these had several sets of magnets and coils whose outputs were all in synchronism. Stöhrer's object was to reduce mechanical vibration, and the electrical output was just as pulsating as the output of a single machine. The credit for first designing a magneto to give a continuous output is due to Wheatstone. However, a machine with five commutators – ten sliding contacts – in series could not really be a practical proposition.

The first commercially used magneto with a continuous output was Woolrich's

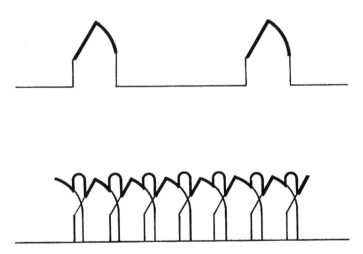

Figure 8.2   The waveform of the current from Wheatstone's simple magneto with a single armature (above) and from his 'magneto-electric battery' (below).

machine of 1844. This had eight coils rotating between the poles of four horseshoe magnets, and the commutator was arranged so that the load circuit was connected only to those coils in which an electromotive force was being generated in the desired direction. This machine was used for electro-plating.

The patent specification of Wheatstone's continuous output magneto is noteworthy for containing one of the earliest descriptions of a complete system of electricity generation, control and application. The control system was the variable resistance or, as Wheatstone called it, the 'rheostat'. The applications suggested include the telegraph, of course, and also driving electric motors and powering electro-chemical processes. He said in the specification that the rheostat was:

applicable to various purposes in which a current of electricity is the agent employed. First, to regulate or modify the velocity of electro-magnetic engines; second, to enable the operator to produce a constant effect in a given time, by retaining the current at the required intensity in the operations of voltaic-typing, voltaic gilding, voltaic decompositions, and other operations requiring regulation. It will in general be useful to interpose in the circuit a galvanometer, in order that the equality, augmentation, or diminution of the current may be indicated and observed. To obtain a constant or nearly a constant current by either of the above regulators during a considerable time, notwithstanding the fluctuations of the battery, all that is necessary is to observe at the commencement of the operation the point at which the needle of the galvanometer stands, and at stated times to bring it back to the same point by means of the regulator, if it should indicate an augmentation or diminution of the current.

Wheatstone made one further contribution to the subject of electrical generation, and then left the subject for over ten years. This was contained in his next patent, obtained jointly with Cooke in 1845, which was mainly concerned with a variety of improvements in telegraphs. However, it also included the idea that a 'voltaic magnet' – or, as we should say, an electromagnet – should be used in a telegraph magneto in place of the permanent magnet. An electromagnet supplied from a battery could be made much stronger than any permanent magnet at that date, so that the output of such a magneto would be correspondingly greater. But this idea does not seem to have been adopted in practice, and permanent magnet generators remained the norm for another 20 years.

Electromagnetic engines

Wheatstone was interested in all the possible applications of electricity. Although he had greatest success with the telegraph, he worked on other applications, especially electric motors. The following note is among his papers at King's College:

This revelation of a new physical power to man which scarcely dates beyond the commencement of this present century, is to exert a most important influence on the future destinies of his race. An energetic source of heat, of light, of chemical action and of mechanical power – already have economical applications of great value been suggested from the development of its laws, which only require to [be] better understood and more determinately refined, to obtain from them those profitable results.

Wheatstone made a number of electric motors or, as they were called at that time, electromagnetic engines. Several machines made within a few years of 1840 are in the Wheatstone Collection in the Science Museum. The earliest has a wooden wheel 35 centimetres in diameter with 16 soft-iron armatures mounted around the circumference. Two horseshoe-shaped electromagnets are mounted just off the horizontal diameter of the wheel so that when one of the armatures is adjacent to one electromagnet the other electromagnet is midway between two armatures. The commutator has 16 conducting and 16 insulating segments of equal size and there are two wipers, one connecting each electromagnet. The arrangement is such that the electromagnets are energized alternately and always with the same polarity. (In some early machines the armatures were permanent magnets which were attracted and then repelled by reversing the polarity of the electromagnets.) Wheatstone may have had his motor made for lecture demonstration purposes after seeing or reading about the machine made by W.H. Taylor.

Taylor's machine (Figure 8.3), which had seven armatures and four electromagnets, was demonstrated in London driving a lathe turning articles of wood, metal and ivory in the early part of 1840. It was the subject of a patent granted in November 1839. In his patent specification Taylor claims as the essence of his invention the feature that the electromagnets are switched on and off by the commutator and never reversed in polarity. An anonymous writer in the *Mechanics' Magazine* of May 1840 wrote eloquently about Taylor's machine and clearly believed in the novelty of Taylor's idea:

The peculiar principle of action, which gives the present invention such a superiority over all previous contrivances, for obtaining a working power from electro-magnetism (the best of them mere toys) is thus explained by Mr. Taylor.

The generality of the plans which have been hitherto devised for the purpose, have depended, he states, and we believe truly, as taking advantage of the *change of polarity*, of which masses of iron fitted as electro-magnets are susceptible so as to cause them alternately to attract and repel certain other electro-magnets brought successively within the sphere of their influence, and thus to produce a continuous rotary movement; and the failure of these attempts he assumed to be owing to the difficulty, if not impossibility of multiplying or accumulating power by such means. Instead of this, Mr. Taylor employs as his prime movers, a series of electro-magnets, 'which are alternately and (almost) instantaneously magnetized and demagnetized without any change of polarity whatever taking place, and in bringing certain other masses of iron or electro-magnets, successively under the influence of the said prime movers when in a magnetized state, and in demagnetizing the said prime movers as soon (or nearly so) and as often as their attractive power ceases to operate with advantage'. Or, 'in other' and perhaps plainer 'words', his invention consists 'in letting on or cutting off a stream of the electric fluid, in such alternate quick and regular succession to and from a series of electromagnets, that they act always attractively or positively only, or with such a preponderance of positive attraction, as to exercise an uniform moving force upon any number of masses of iron or magnets, placed so as to be conveniently acted upon.

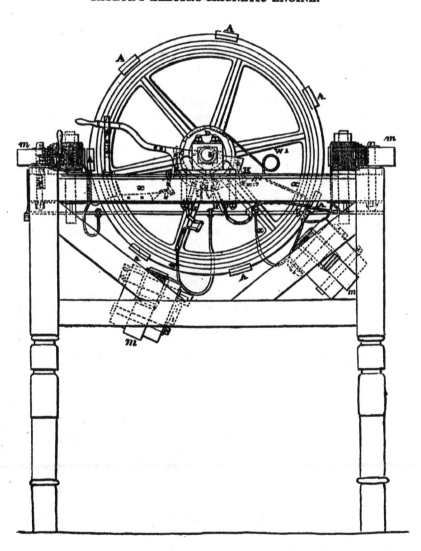

𝕸𝖊𝖈𝖍𝖆𝖓𝖎𝖈𝖘' 𝕸𝖆𝖌𝖆𝖟𝖎𝖓𝖊,

MUSEUM, REGISTER, JOURNAL, AND GAZETTE

| No. 874.] | SATURDAY, MAY 9, 1840. | [Price 3*d.* |

Printed and Published for the Proprietor, by W. A. Robertson, No. 166, Fleet-street.

TAYLOR'S ELECTRO-MAGNETIC ENGINE.

Figure 8.3   W.H. Taylor's electromagnetic engine which was demonstrated near King's College in 1840, and featured in the Mechanics' Magazine. *Wheatstone and his engineering students must have seen it.*

Taylor's machine was probably the first Wheatstone had seen which was capable of doing useful work. The idea of simply switching the magnets on and off instead of reversing their connections made his machine far simpler than Sturgeon's motor of 1832 (which Wheatstone would have known about) with its complicated arrangement of mercury commutators for changing the connections of the coils.

In the few months preceding its account of Taylor's engine the *Mechanics' Magazine* published reports of electromagnetic engines being used by Professor Jacobi in Russia to drive a boat carrying ten people on the River Neva, and by Thomas Davenport in New York to drive a printing press requiring two horsepower. These machines, like Sturgeon's, operated by reversing the electromagnets. The *Mechanics' Magazine* also reprinted in November 1839 the observation by Dr Jones, Editor of the *Journal of the Franklin Institute* that 'it has been satisfactorily ascertained by numerous and well-conducted experiments . . . that . . . motive power by the electro-magnetic influence is a thing not to be hoped for, in the present state of our knowledge upon that subject'; but in June 1840 it was reporting that the Scot Robert Davidson had made an electrically-driven railway vehicle. In 1842 Davidson conducted a demonstration on the Edinburgh and Glasgow railway. His motors, like Taylor's, worked by switching the magnets on and off.

Despite the enthusiastic manner in which the *Mechanics' Magazine* acclaimed his machine, Taylor's idea was not actually novel. The writer had not looked carefully enough through the back numbers of his own journal! In 1833 it had published a letter giving a description – admittedly a somewhat sketchy one – of an electromagnetic engine which stated explicitly that the connections to the electromagnet were to be periodically interrupted or reversed.

Against this background it would not be surprising if Wheatstone wanted an electromagnetic engine to try out himself. Possibly the machine which now survives was made in the College for demonstrating to students, or even made by the engineering students themselves. This is purely speculation, but we have seen that they were expected to exercise their 'observation, judgement, and invention' by 'experiments made by themselves' and to make industrial visits with their teachers. In the spring of 1840 the College's first engineering class were approaching the end of their first year, and as part of the course on electricity Wheatstone might well have taken his class to see Taylor's machine, which was exhibited less than a mile away at the Coliseum. After such a visit what better 'experiment' for the students than to make an engine of their own? The machine is crudely constructed and mostly of wood (see Figure 8.4), whereas most of Wheatstone's experimental apparatus is of metal and made to a high standard.

Three other motors made by Wheatstone and now in the Science Museum are quite different from the machine just described. They represent an attempt to progress from the lecture demonstration to a practical power source. Wheatstone told Grove that he called these his 'eccentric electro-magnetic engines', but the term is not used in any printed account.

One basic problem with all the early designs of electromagnetic engines arises

Figure 8.4   Wheatstone's first electromagnetic engine which may well have been inspired by seeing Taylor's machine. It is now in the Science Museum, and will run if connected to a 12 volt battery.

from the fact that while the force between an electromagnet and a soft-iron armature may be considerable when they are almost in contact, it falls off rapidly with increasing distance. Although a great deal of power may be available during a very short range of travel, it is difficult to make practical use of this short movement. Wheatstone's solution was to lengthen the working stroke by making the armature move in a direction inclined to the direction of the magnetic attraction, so that the driving force was applied over a greater distance than if the armature moved directly along the line of force. He was not thinking that an increased power output would be obtained by making the driving force active over the longer distance. The relationship 'work done equals force multiplied by distance moved in the line of force' was already well known. Wheatstone's friend and colleague Professor Daniell described two of these machines in the 1843 edition of his textbook of chemistry. He said the virtue of the new principle Wheatstone had introduced into the construction of electromagnetic engines was that 'the armatures are constantly, during the existence of the attraction, in presence of the magnets, and in very close proximity with them, whereby the motion is rendered more energetic and less abrupt'.

The machines are described in the specification of Wheatstone's patent

No. 9022 of 1841, though the specification does not indicate why the design was thought to be advantageous. The description of the principle of working is given in purely geometrical terms. Essentially it says that whereas other electro-magnetic engines have electromagnets and armatures with their active faces arranged in concentric circles (as is the case also with modern rotating electrical machines) these engines have their armature surface in a circle which is not concentric with the circle linking the surfaces of the electromagnets. The two circles could have either parallel axes or intersecting nonparallel axes.

The first two extant eccentric machines described below have parallel axes, and the third has intersecting non-parallel axes.

The eccentric machines were lent by Wheatstone to William Grove to illustrate a lecture in 1844. Among the Grove papers at the Royal Institution, where he was a Professor, are letters from Wheatstone including several responding to requests to borrow apparatus for lectures, though there is no account of a lecture using Wheatstone's machines. On this occasion Wheatstone readily agreed to the loan but told Grove that he had not published anything about the principle of the machines. He thought a proper account would be helpful, for, as he remarked to Grove, 'I cannot expect that you will be able to give so good an account of them as I could have enabled you to do after an hour's explanation'. Wheatstone never did publish an account of the principle of these machines. The Patent Specification describing the machines was not published until after the reform of the patent system in 1852, and that is purely a mechanical description of the machines rather than an explanation of how they work.

The first of the eccentric engines (see Figure 8.5) was said by Wheatstone to be the least useful of the three but to explain the principle best. It has eight horseshoe electromagnets supported inside a brass ring 25 centimetres in diameter. The armature is a soft-iron ring of half that size, mounted on a crank with a throw of one centimetre. A static eight-part commutator is fixed to one bearing of the main axis and a single wiper rotates with the shaft. Each of the eight segments of the commutator is connected to one of the horseshoe magnets; one pole of the supply is connected to the common terminal of all the electro-magnets and the other is connected to the wiper through the bearings and frame. The arrangement is such that when the armature is closest to one of the horse-shoe magnets the commutator connects the next magnet in sequence so that the armature rolls around the inner surfaces of the magnets pulling the crank around with it.

The description of the first eccentric engine in the patent specification differs in one major respect from the machine now in the Science Museum (see Figure 8.6). According to the specification there is no commutator on the shaft but the inner surfaces of the magnets are bounded by an additional ring carrying eight springy metallic contacts; as the armature rolls around it touches each of these contacts in turn. The machine now extant does not appear to have been altered, and there is no evidence that any machine with the contact arrangement described in the specification was ever built. The description in Daniell's book is similar to that in the specification, and he implies that the machine had actually

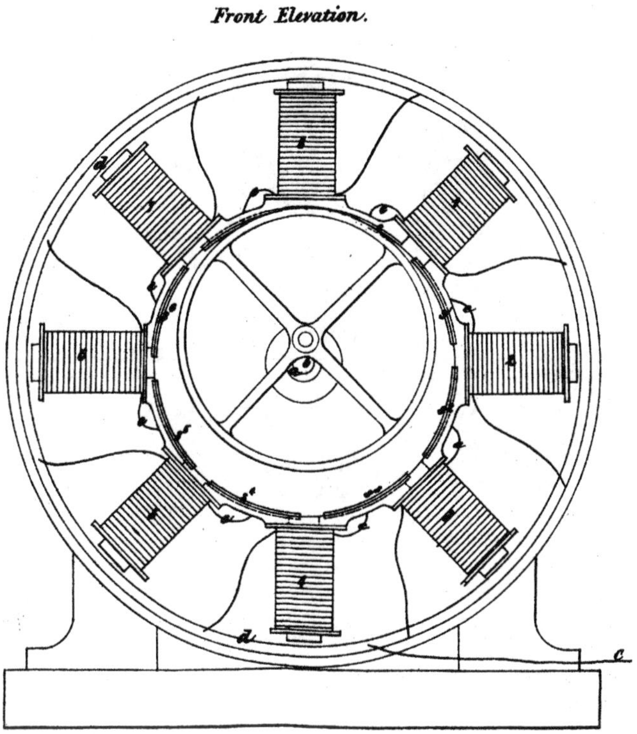

Front Elevation.

Figure 8.5   *The patent drawing of Wheatstone's first eccentric electromagnetic engine.*
The eight horseshoe electromagnets are mounted in a brass ring and their poles
form a smooth circular surface. The armature is a soft iron ring mounted on a
crank. The magnets are energised sequentially, through the contacts which can
be seen at the inner ends of the electromagnets, and the armature rolls round
pulling the crank with it.

been constructed. Possibly Daniell copied the description from the specification
and saw the machine in Wheatstone's laboratory, but did not notice the
discrepancy.

The second machine (see Figure 8.7) has two fixed horseshoe magnets inclined
at 45° relative to one another. The rotor is a wooden disc about 12 centimetres in diameter carrying four soft-iron arcs which are not concentric with the
axis. The machine has an eight-part commutator on the shaft similar to the
commutator of the preceding machine but alternate segments are connected
together and to one of the electromagnets so that in each successive quarter of a
revolution first one magnet and then the other acts on one of the four arcs.
There is no description corresponding to this machine in the patent specification, but Daniell refers to it and also to another machine then being constructed
which would be of much greater power and 'which will have, in the same space,
four times the power of either of the preceding arrangements'. Probably this

Figure 8.6   Wheatstone's first 'eccentric' electromagnetic engine.

Figure 8.7   Wheatstone's second 'eccentric' electromagnetic engine.

machine would have combined the stator of the first machine with the rotor of the second. Such a machine, however, would not have worked very well and would not have had the power hoped for. In both of the first two machines only one electromagnet acts at any one time. In the hypothetical machine the alternate electromagnets (1, 3, 5, etc.) would have been energized simultaneously. An armature which had been brought close to electromagnet 2 would, when the wipers passed from one commutator segment to the next, be pulled forward by electromagnet 3 and retarded by electromagnet 1. Although the attraction of 3 would be greater and the machine would move, the total power would be less than four times that of the simpler machine.

The third extant machine (see Figure 8.8) was referred to by Wheatstone as his 'eccentric disc machine'. It is not mentioned in Daniell's book. The armature is a soft-iron disc fixed on an axis inclined to the main axis of the machine. One end of the inclined axis is pivoted on the centre line of the machine, and the other end is pivoted on a crank on the main axis. The disc is free to rotate. Four horseshoe electromagnets are disposed around the edge of the disc on one side of it. As each electromagnet is energized in turn, through the four-part commutator on the main shaft, the disc performs a wobbling motion and the inclined axis describes the surface of a cone, causing the crank and the main axis to rotate. This machine alone of the three is provided with a pulley from which a drive might have been taken, though there is nothing to indicate what output, if any, was ever obtained from it.

The eccentric disc machine bears a remarkable similarity to an unusual design of steam engine which was developed in the 1830s. This was the disc engine, patented in 1830 by E. and J. Dakeyne and then developed by a number of other workers. The disc engine has a barrel-shaped cylinder and its 'piston' is a disc fixed on a ball and socket joint at the centre of the barrel. The disc performs the

Figure 8.8   Wheatstone's third 'eccentric' electromagnetic engine.

same wobbling motion as the armature of Wheatstone's eccentric disc machine and is coupled to the output shaft by a similar connecting linkage. In both machines the disc is driven by a force whose centre of application moves continuously around it. Many early electromagnetic engines were consciously based on steam engine designs, and it is probable that Wheatstone was intending to make an electrical machine analogous to a steam engine whose operating principle had not been tried out electrically.

The drawing of this machine in the patent specification (Figure 8.9) is quite elaborate and exhibits one strikingly modern feature. In place of the four horseshoe magnets clamped on a wooden frame in the surviving machine, the patent drawing shows an iron disc with grooves cut in it to accommodate the windings.

Only one other person appears to have made an eccentric electromagnetic engine, and that was the Frenchman Froment who constructed a number of engines (not all eccentric) between 1844 and 1848. Froment's machines were described by du Moncel in an article in 1883, and one eccentric engine is exhibited at the present time in the Musée du Conservatoire des Arts et Métiers in Paris. According to du Moncel, Froment made his eccentric machine, which he called 'électro-moteur epicycloïdal' in 1847, and Wheatstone's priority is clearly stated.

The idea that the working stroke of an electromagnetic engine might be increased advantageously was taken up by another inventor, Thomas Allen. In 1852 he developed a machine similar to a four-cylinder steam engine in which each piston and cylinder was replaced by four electromagnets and armatures. The armatures could slide along the 'piston rods', but pressed on collars fixed to

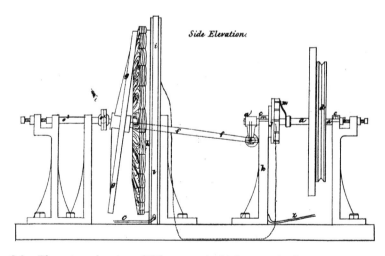

Side Elevation.

*Figure 8.9   The patent drawing of Wheatstone's third eccentric electromagnetic engine. The horseshoe electromagnets **h** are mounted around a slightly coned surface. As the magnets are energised sequentially the soft iron disc **g** performs a rolling motion around the magnets so that its shaft **f** drives the output shaft through a crank.*

the rods. In each power stroke the first armature pulled its rod through one quarter of its travel; the armature then came to rest on the face of its electromagnet but the rod continued moving under the action of the second armature and electromagnet which were effective for the second quarter, and so on. A number of machines of this design were made, and there is one in the Science Museum.

Wheatstone's three eccentric electromagnetic engines languished in well-deserved obscurity for more than a century until they were literally unveiled before an audience of leading electrical engineers. The occasion was the inaugural address given by the late Professor E.R. Laithwaite when he became Chairman of the Power Division of the Institution of Electrical Engineers in 1971. He chose for his theme 'The shape of things to come', looking at the shape of objects as diverse as teapots and butterflies in relation to their function and then turning to the design of electrical machines. In preparing the address he studied the collection of early electric motors and generators in the Science Museum to see the many varieties of design that have been tried out in the past. The present author showed him the eccentric engines, which were not on exhibition, and he was so fascinated by them that they were promoted from the Museum Store to pride of place in the exhibition of unusual machines that accompanied his address.

Linear motors

Professor Laithwaite was famed for his pioneering work on the development and application of linear electric motors. At the very time he was addressing the Institution and paying tribute to Wheatstone's ingenuity there was, lying in a cupboard a few hundred yards away in King's College, a linear motor made by Wheatstone in the 1840s. Wheatstone's patent specification mentioned above, which describes the eccentric electromagnetic engines, also includes the earliest description of a linear motor. After stating that in the eccentric engines the electromagnets and armatures have their active faces arranged in non-concentric circles, the specification goes on to state that one of the circles might be unrolled into a straight line in which case the other circle would roll along it and produce a reciprocating machine. Today linear motors are often described as 'ordinary' motors which have been slit longitudinally and then unrolled; it is interesting that Wheatstone described his in the same way:

Instead of two curves, as above mentioned, a curve and a straight line may be employed, the electro-magnets being arranged in a straight line and the armatures in a curve (or vice versa). In this case a reciprocating motion is obtained, and not a motion of revolution, the curve rolling on the straight line until it reaches the end of it, then rolls back again in the contrary direction, to its other end.

The mere existence of a description in a patent specification is no proof that the device was ever made. The absence of any reference to it in the 1843 edition of Daniell's book mentioned above seems to imply that it had not been made.

Figure 8.10   The stator of Wheatstone's linear motor.

But one was made (see Figure 8.10). In the Physics Department of King's College London there was a cupboard containing Wheatstone relics together with a quantity of later material which someone, presumably, was reluctant to throw away. I was permitted to search through this cupboard in 1973 and found an item of particular interest literally buried under broken telegraphs, concertinas and other things. This is a piece of iron 30 centimetres long by 12 wide by 2.5 thick, with slots cut into it leaving 12 'teeth'. The slots are filled with windings of thick wire, about 12 turns around each tooth, and the whole is mounted on a wooden baseboard. The ends of the windings are brought out to 13 brass inserts in a strip of wood fixed to the baseboard, and the arrangement is such that the winding around each tooth is connected between two adjacent brass pieces. It had been overlooked since Wheatstone's time, but having recently read the patent specification I was able to recognize it as the stator of a linear motor. Professor Laithwaite agreed with my conclusion and we decided to try out the newly found machine. Wheatstone's rotor and brushgear did not survive, so we experimented to find what was best.

A convenient rotor was found to be a piece of steel tube about five centimetres in diameter and as long as the teeth. Current was supplied through two carbon brushes held by hand on the brass pieces. When the brushes were moved from one end of the machine to the other the rotor rolled along too. For best results the brushes were spaced two or three teeth apart (so that two or three coils were energized simultaneously) and a piece of thin card was laid on the top of the teeth to give a smooth path for the rotor.

Why was nothing further heard of Wheatstone's linear motor? The experiments just described give the answer. The machine works, but requires a current of 30 Ampères for the rotor to move at all, and for reliable operation at least 50 Ampères is required. Even higher currents would be needed for the machine to do useful work, and it is unlikely that Wheatstone could have obtained enough current to demonstrate the machine. The brass contact pieces show no signs of burns from sparking, which would undoubtedly be present if it had been much used.

It is of interest to note that W.H. Fox Talbot obtained a patent in 1852 for an electromagnetic engine with a linear motion.[7] According to the patent specification his machine had a row of horseshoe electromagnets and an iron cylinder rolling across them as they were energized successively. Wheatstone did not suggest any application for his linear machines. Talbot described a crank linkage to convert the reciprocating motion to rotary. There is no record of anyone else making a linear motor until the end of the century. The first practical use of a linear motor was for an aircraft catapult during the Second World War.

Although he failed to produce an electric motor as a practical source of power, Wheatstone's work in this field cannot be dismissed as a complete failure. The receivers of his 'ABC Telegraphs', which will be considered later, all contain a tiny motor (typically a few centimetres across) which drives the pointer, and he made electric clocks operating in the same way. With a modern understanding of machine theory it can be shown that Wheatstone's smaller motors are bound to work better than the large ones. They are all of the class of machine now known as 'reluctance' machines. In a reluctance motor the moving piece – the armature or rotor – tries to move into such a position that the reluctance of the magnetic circuit is at a minimum. In other words, it is a purely magnetic machine operating by magnetic attraction across the air gap, and not an electromagnetic machine depending on interaction between the magnetic flux in the air gap and the current in the armature.

Most electric motors and generators are electromagnetic machines and their efficiency increases with size. Essentially the reason is that the power output is proportional both to the magnetic flux in the air gap and to the current induced in the armature. Both these quantities are proportional to the square of the linear dimensions of the machine, and the power output is therefore proportional to the fourth power of the linear dimensions. The weight of the machine, however, is proportional to the cube of the dimensions and the power to weight ratio therefore increases with size.

In a magnetic machine the considerations are different. The magnetic force between rotor and stator is proportional to the square of the magnetic flux, and therefore may be thought to be proportional to the fourth power of the linear dimensions. But in all Wheatstone's designs of motor the flux path through the air increases in proportion to the linear dimensions and, since the maximum magneto-motive force is determined by the available magnetic materials, the flux density decreases in proportion to the increase in linear dimensions. The net result therefore is that the power output increases in proportion to the square of the linear dimensions. Since the weight increases as the cube of the linear dimensions, the power to weight ratio decreases as the size of the machine increases.

Wheatstone would not have had the theoretical understanding to know why the larger motors did not prove practicable power sources, but he was a practical engineer as well as an academic scientist and, having failed to make a useful power source despite many ingenious ideas, he directed his efforts to further refinements of the telegraph motor which he subsequently brought to an advanced state of perfection.

He retained the hope that the electromagnetic engine would one day become a practical source of mechanical power. In his Royal Society paper on measurements in 1843 he describes the rheostat and says that it can also be used for controlling the speed of a motor, or keeping it constant as the battery varies. He also stated:

Since the consumption of materials in a voltaic battery in which there is no local action decreases in the same proportion as the increase of the resistance in the circuit, this method of altering the velocity has an advantage which no other possesses, the effective force is always strictly proportional to the quantity of materials consumed in producing the power, a point which, if further improvements should ever render the electro-magnetic engine an available source of mechanical power, will be of considerable importance.

This, of course, is not correct, since it ignores the power wasted in the resistance, but it suggests that he had conducted some experiments designed to relate the power obtained from an engine and the consumption of materials in the battery.

When the first edition of this book was published the story of reluctance motors ended there. Such motors were hardly used except in mains-frequency clocks. In that application the power is tiny and the necessary switching is performed automatically by the alternations of the supply. Wheatstone only had a direct current supply from a battery, and his motors had commutators with sliding contacts to do the switching. Wheatstone's machines did work, and those in the Science Museum still work if connected to a suitable battery (six or twelve volts). When running there is considerable sparking at the commutator, and clearly the brushes – small strips of copper – would not have lasted long. There is very little sign of burning, so we may assume that Wheatstone never ran them for long.

Since about 1980 switched reluctance motors have been again coming to the fore, largely due to the work of Professor Peter Lawrenson of Leeds. Better designs and the introduction of improved magnetic materials have both contributed to the new developments, but the most significant factor has been the introduction of power semiconductor devices which enable quite high currents to be switched rapidly, at precisely controlled times, and with no damaging sparking. Such technology was not available to Wheatstone, but he was responsible for some of the early research from which modern electric drives have developed and which are used in such diverse applications as domestic electrical appliances and main-line railway trains.[8]

Detonating explosives: induction generators

Another application of electricity which interested Wheatstone was the detonation of explosives. About 1840 Colonel Sir Charles Pasley was experimenting with underwater explosives and, according to W.T. Jeans in his *Lives of the Electricians* published in 1887, Pasley consulted Wheatstone and Daniell on the question of electrical detonation. There is no contemporary record of Wheatstone's involvement, but Jeans stated that they conducted experiments at King's College to determine the best procedure, and that Pasley made use of their advice when removing the wreck of the *Royal George* at Spithead.[9]

The *Royal George*, flagship of Admiral Kempenfelt and pride of the Royal Navy, had sunk while at anchor at Spithead on 29 August 1782. The ship had been deliberately heeled over to enable carpenters to carry out repairs, when a sudden breeze caused her to sink in a few minutes with the loss of nearly 1,000 lives. Several attempts were made to raise the vessel, both for salvage and because the wreck was a danger to other shipping, but little success was achieved until Pasley commenced operations in 1839. He had previously blown up some old wrecks in the Thames to clear a channel, and his plan was to break up the *Royal George* and salvage as much as possible with the aid of divers. In his first season of work, the summer of 1839, Pasley detonated his charges with chemical fuses, but in the next few seasons he successfully introduced an electrical ignition system in which an electrically heated platinum wire detonated the explosive charges. An account published in 1842 states that Pasley used various different batteries including Daniell's. Since Wheatstone and Daniell were close friends and colleagues it seems probable that Wheatstone himself would have been involved in the experiments which led Pasley to use electrical ignition. Fifteen years later, as a member of the Select Committee on Ordnance, Wheatstone was certainly involved very actively in experiments of that nature.

Notes

1 For a general account of the background to this chapter see Percy Dunsheath, *History of Electrical Engineering*, 1962; Brian Bowers, *A History of Electric Light and Power*, 1984 and second edition in preparation.
2 Wheatstone referred to a paper by Linari in *L'indicatore Sanese*, No 50, 13 December 1836.
3 I am grateful to Dr Bernard S. Finn of the Smithsonian Institution, Washington, for the extract from Henry's *Journal*.
4 Faraday's experiments with Wheatstone at the Adelaide Gallery were recorded in Faraday's *Diary*, which was published by the Royal Institution in 1934. The quotation is from 26 November 1838.
5 Cooke's letters are in the Archives of the Institution of Electrical Engineers. Some, including this one, have been published in F.H. Webb (ed.) *Extracts from the private letters of the late Sir William Fothergill Cooke, 1836–39,*

relating to the Invention and Development of the Electric Telegraph, London, 1895.

6   Th. du Moncel, *Exposé des Applications de l'Electricité*, tome 2, Paris, 1873.

7   English Patent No. 1046 of 1852.

8   Brian Bowers, 'Switched Reluctance Motors – an old technology reborn', *Engineering Design, Education & Training*, Design Council, Spring 1989, pp. 28–31. The website of Switched Reluctance Drives Ltd also has an historical introduction referring to Wheatstone's involvement.

9   For an account of the sinking and salvaging of the *Royal George* see the booklet by 'J.S.', *A Narrative of the Loss of the Royal George at Spithead . . .*, Portsea, 1842.

Chapter 9

Electrical measurements

The name of Wheatstone is always associated with the bridge circuit for measurements, notwithstanding the fact that he called it 'The Differential Resistance Measurer' and explicitly gave the credit to S.H. Christie. He gave an account of it in 1843 when he was invited to give the annual Bakerian Lecture to the Royal Society; his title was 'An account of several new Instruments and Processes for determining the Constants of a Voltaic Circuit'. The term 'bridge' was not used in the published paper.

A writer in 1931 said the bridge circuit was so called from its 'resemblance in the early sketches to the Chinese bridges familiar in the willow pattern'. I have not found this idea in any earlier account, and the reader is invited to judge its likelihood from the drawings (see Figure 9.1). Another, perhaps more plausible, suggestion is that the term 'bridge' relates not to the whole circuit, but to the detector, which 'bridges' points of equal potential.[1]

Wheatstone had been working on the problems of electrical measurements at least since 1840, but it is probable that the work began with his early telegraphic experiments, possibly as early as 1836. This is supported by the reason Wheatstone gave for studying measurements. In the first section of the 1843 paper he said:

The practical object to which my attention has been principally directed, and for which these instruments were originally constructed, was to ascertain the most advantageous conditions for the production of electric effects through circuits of great extent, in order to determine the practicability of communicating signals by means of electric currents to more considerable distances than had hitherto been attempted.

Jacobi said in 1840 that Wheatstone was principally concerned with the measurement of electromotive force.[2] By that date he had ascertained 'the most advantageous conditions for the production of electric effects' to his own satisfaction, and was making measurements on the various sources of electricity which he was investigating about that time. After stating his object in the paper, Wheatstone added 'guided by the theory of Ohm, and assisted by the instruments I am about to describe, I have completely succeeded'. Ohm's study of the

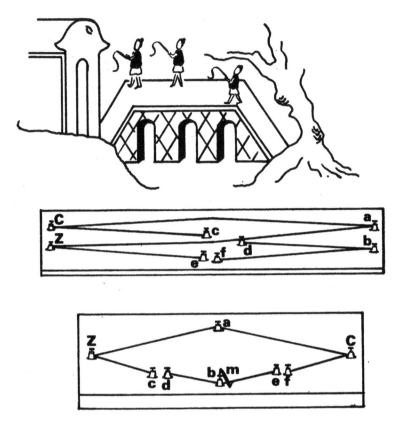

Figure 9.1   Two bridge drawings from Wheatstone's measurements paper, and (above) the Chinese bridge in the willow pattern.

mathematics of electric circuits had been published in Germany in 1827 but was virtually unknown elsewhere.[3] It was highly relevant to the electric telegraph, but not to the only earlier applications of electricity, electrochemistry and the arc, whose circuits do not obey Ohm's Law. Wheatstone had read Ohm's work and was in a position to have it brought to the knowledge of English-speaking scientists. He was a member of the committee appointed by the British Association in 1838 to 'superintend the Translation and Publication of Foreign Scientific Memoirs'. The committee had £100 per year at its disposal and it arranged for translations to be prepared and published in *Taylor's Scientific Memoirs*. It is a measure of the importance of Ohm's work that his paper, originally published over a decade previously, was the first on an electrical subject to be translated and published in English under the auspices of the committee. It appeared in 1841.

Early measuring techniques

Before looking at Wheatstone's paper on measurements, there is an earlier idea of his which was not included in the formal paper but deserves notice. It is recorded only on a scrap of paper among his notes at King's College and concerns the measurement of very small electromotive forces. To measure an electromotive force which is so weak that it produces a barely perceptible deflection of the galvanometer needle the wire is touched on the galvanometer terminal briefly and then removed. The galvanometer needle will receive a brief impulse so that it swings forward a little, then back in the opposite direction, then forward again, and so on. Wheatstone's method was to touch the wire on the galvanometer terminal again as the needle was moving forward so that it received another brief impulse and increased its amplitude of movement. By repeating the process the amplitude could be increased until the needle hit the end stop of the instrument. The number of impulses needed to achieve this would be a measure of magnitude of the electromotive force.

The galvanometers available at the time would have oscillated readily. They consisted of a coil wound on a non-magnetic former and a pivoted or suspended magnetic needle. Iron was not introduced into the core of a galvanometer until 1845, when Wheatstone patented the idea. The galvanometer itself had been known since 1821, when Schweigger repeated Örsted's experiment but used a coil of wire instead of a single wire near a compass needle.

The simple galvanometer will indicate the presence of a current and it will show that one current is greater or smaller than another. But it cannot indicate numerical relationships between the currents until the relationship between the current and the angle of deflection is known. Though several workers, including Wheatstone, sought to 'calibrate' a galvanometer so that the magnitude of a current could be inferred from the angular deflection of the needle, Wheatstone's great achievement in the field of electrical measurements was to show how measurements could be made without any need for a calibrated galvanometer.

He devised circuits using a calibrated variable resistance. The function of the galvanometer was no longer to measure current but simply to show that the current observed with two slightly different circuits had the same magnitude. A calibrated variable resistance was easy to make because Ohm's work showed that the resistance of a conductor of uniform section was proportional to its length. For a standard of resistance Wheatstone took the resistance of a copper wire one foot (300 millimetres) long and weighing 100 grains (6.48 grams). This corresponds to a diameter of 1.8 millimetres, but he pointed out that greater accuracy can be obtained by measuring length and weight than by measuring the diameter. The unit is about one fifth of an Ohm.

An example of earlier electrical measurement techniques is the method of measuring resistance used by the German physicist Fechner.[4] In 1831 he showed how to determine the resistance (R) of a circuit by measuring the current (F) first with, and then without, a known additional resistance (r) included in the circuit. This gives two equations, which can be solved for (R) provided that the electromotive force is constant:

For the circuit without additional resistance

$$F = E/R \qquad\qquad \text{Equation 1}$$

For the circuit with additional resistance

$$F' = E/(R + r) \qquad\qquad \text{Equation 2}$$

Solving these equations gives

$$F/F' = (R + r)/R$$

from which R can be calculated.

The principle is very simple; the problem is the measurement of F and F'. Fechner measured the force of the current by the very tedious process of observing and timing the oscillation of a magnetized needle at right angles to a coil carrying the current. Other workers had measured the current by observing the deviation of the needle, and then inferring the corresponding degrees of force by what Wheatstone called 'some peculiar process'. One of the people who deduced the magnitude of a current from the deflection of a magnetic needle was Pouillet, who in 1837 described the tangent and sine galvanometers and showed that they gave an accurate measure of current.[5] He called the instruments the 'compass of tangents' and the 'compass of sines', although the term 'galvanometer' was already in use, and he showed that the tangent galvanometer could give results consistently within ± 4 per cent of the mean, so that Wheatstone's remark just quoted seems rather harsh. However, Wheatstone also pointed out that even if a galvanometer were calibrated to measure the force of the current it would not be very reliable over a period of time because of the variations in the magnetization of the needle.

The Bakerian Lecture

Wheatstone began his Bakerian Lecture with some definitions of terms and a short account of the principal results which derive from what he referred to as Ohm's 'beautiful and comprehensive theory'. If an *Electromotive force* (*E*) is the cause which originates an electric current in a closed circuit or gives rise to an 'electroscopic tension' in an unclosed circuit, and *Resistance* (*R*) signifies the obstacle to the passage of the current and is the inverse of the conducting force then, according to Ohm, the *Force* of the current (*F*) is given by

$$F = E/R$$

The length of copper wire of a given thickness having the same resistance as a circuit was called by Ohm the Reduced Length of that circuit. Wheatstone then derived a number of equations relating to simple circuits, especially circuits with resistances in parallel. He based his discussion on the use of both voltaic and thermo-electric elements, which have relatively high and low internal resistance respectively.

He thought it necessary to introduce a number of additional technical terms, since 'it is seldom that any real advance is made in a scientific theory without a corresponding change in its terminology'. He said that it was then proved beyond doubt that the various sources of continued electric action differ only in the amount of their electromotive forces and their internal resistances (he had not been certain about this in 1837 when writing on thermo-electricity). The new terms he used were all based on the Greek root *Rheo*, meaning a stream or current:

- *Rheomotor* any apparatus which originates an electric current.
- *Rheometer* an instrument to measure the force of an electric current.
- *Rheotome* an instrument which periodically interrupts a current.
- *Rheotrope* an instrument which alternately inverts a current.
- *Rheoscope* an instrument which merely ascertains the existence of a current.
- *Rheostat* (the only word to survive in modern usage) which Wheatstone originally defined as a device for adjusting the resistance of a circuit so as to keep the current constant. Wheatstone himself extended the meaning of the word to include any variable resistance, whatever its function.

Wheatstone continued to use the established terms *galvanometer* and *voltameter* for specific instruments, but *rheometer* was a general term for any electrical measuring instrument. He justified his new terminology by pointing out that Ampère had used the term *Rheophore* for a connecting wire and that *Rheometer* had been generally adopted by French writers on physics as a synonym for galvanometer.

The terms rheometer, rheomotor and rheostat are used throughout the paper. Rheometer is used again in the paper written by Wheatstone in 1860 for the committee enquiring into the construction of submarine telegraph cables, and rheomotor was used by the Electrical Standards Committee of the British Association in 1862. Rheotome was used in the 1870s by Alexander Graham Bell, but rheotrope and rheoscope do not seem to have been used again.

The basis of Wheatstone's method of measuring circuit constants is the use of variable resistance to keep the current constant and so avoid any need for calibration of the galvanometer. In order to utilize this principle it is necessary to have a calibrated variable resistance. He described two rheostats for the purpose, one for high resistances and one for low resistances. The high-resistance rheostat has two parallel cylinders: one wooden with a spiral groove cut in its surface, and the other of brass. In use a wire is wound from one cylinder to the other, and the resistance in circuit is the length of wire wound on the wooden cylinder at any one time. Connections are made through wipers pressing on to a brass ring on each cylinder. The number of complete turns on the wooden cylinder can be read off a central scale, and fractions of turns are indicated by a pointer on the end of the wooden cylinder.

A rheostat which is probably the original high-resistance instrument is now in the Science Museum (see Figure 9.2). There are 190 turns of wire and the scale on the end of one cylinder has 100 divisions. The total resistance is 60 Ohms, so the resistance per turn is about 0.31 Ohm.

To supplement the rheostats and extend the range of resistance measurements

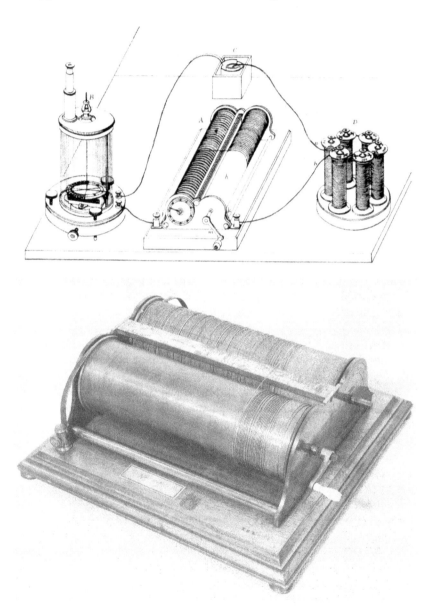

Figure 9.2   *Wheatstone's Rheostat, and drawing from his 1843 paper on measurements showing it in use.*

he could make, Wheatstone also had a set of six resistance coils wound with copper wire 0.005 inches (0.13 millimetres) in diameter. (When discussing these matters in his paper Wheatstone used several different units – the inch, the foot, the yard and the mile.)

He does not describe the galvanometer he uses except to say that it has an astatic needle and a microscope to facilitate reading the scale. Two years later, however, he patented an iron-cored galvanometer (see Figure 9.3).

Wheatstone's paper gives many examples of measurements of resistance, electromotive force and current, and a selection of these will now be described to illustrate his methods. They will sound elementary to a modern reader with any knowledge of electrical engineering, but it should be remembered that Wheatstone was expounding some of the basic principles of the subject for the first time.

He began by showing how to determine the value of an unknown resistance by a simple substitution method. The unknown resistance is included in a circuit with a galvanometer whose reading is noted, and it is then replaced by a rheostat. The rheostat (or rheostat and additional coils if necessary) is adjusted so that the galvanometer reading is the same as before. The scale on the rheostat gives its resistance, and hence the value of the unknown resistance.

He also showed how the resistance of a galvanometer could be found, without calling for another galvanometer, provided that two rheomotors of identical electromotive force and internal resistance were available. One rheomotor of electromotive force E and resistance R is connected by wires of resistance r to the galvanometer whose unknown resistance is g and the galvanometer reading is noted. The second rheomotor and a rheostat are then included in the circuit,

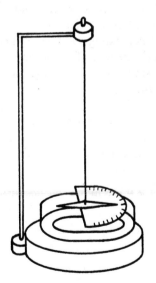

Figure 9.3   Wheatstone's iron-cored galvanometer.
(Redrawn from figure Y of his Patent Specification 10,655 of 1845.)

and the resistance x of the rheostat adjusted so that the galvanometer returns to its first reading. Since the currents are then the same in both circuits

$$E/(R + r + g) = 2E/(2R + r + g + x)$$

whence $g = x - r$.

The measurement of the resistance of a liquid presents special problems. With most conducting liquids some electro-chemical action takes place while the measurements are being made. Consequently there is a voltage drop at the electrodes and the current is not directly proportional to the applied electromotive force. Furthermore there may be changes in the composition of the liquid and the electrodes, and gas may be discharged on the surface of the electrodes which increases the apparent resistance.

No satisfactory measurements of liquid resistances had been made before Wheatstone approached the problem. He placed the liquid in a horizontal glass tube which had been ground away on one side to form a trough. One end of the tube was closed with a platinum plate; the other end was closed with a moveable piston which also had a platinum plate on its face. The range of movement of the piston was exactly one inch (25.4 millimetres) and the spacing between the platinum plates could be varied between ¼ inch and 1¼ inches. The device was then connected in series with a galvanometer, a rheostat, resistance coils if necessary, and a battery consisting of several elements. The rheostat was adjusted to bring the galvanometer needle to a specified point, first with the piston ¼ inch from the end-plate and then with it 1¼ inches away. The difference between the two rheostat settings was then the resistance of a column of liquid one inch long and of known cross-section, and the problems arising from electro-chemical effects at the electrodes were avoided.

Wheatstone then considered the problem of measuring the total electromotive force in a circuit without depending on a calibrated rheometer and without having recourse to Fechner's method of timing the oscillations of a magnetized needle; he thought that a simple method of measuring the electromotive force would be of great value in the study of electro-chemistry. His method utilizes the rheostat as a calibrated variable resistance and a galvanometer to show that equal currents flow in two different circuits. Since the currents are equal, the ratio of the sum of the electromotive forces to the sum of the resistances is the same in both circuits. If n is the ratio of the electromotive forces (and of the resistances) in the two circuits, then

$$F = E/R \text{ for the first circuit}$$
$$\text{and } F = nE/nR \text{ for the second circuit.}$$

Let an additional known resistance r be inserted in the first circuit. The current then falls to F':

$$F' = E/(R + r)$$

To reduce the current in the second circuit to F' it is necessary to insert an additional resistance *nr*, so that

$$F' = nE/(nR + nr)$$

The resistance r is known and nr is found by experiment; hence n is found.

Wheatstone's experimental procedure was to insert the rheostat and galvanometer in the circuit, adjust the rheostat so that the needle of the galvanometer was deflected 45° and then adjust it again until the deflection was 40°. The difference in the two rheostat settings is a measure of the electromotive force in the circuit. Once the difference corresponding to a standard source has been found, the electromotive force of any other source can be calculated in terms of the standard. He gave examples of measuring the electromotive forces of a number of voltaic and thermo-electric elements. He also measured the 'contrary electro-motive force' of a voltammeter or electrolytic cell by finding the reduction in apparent total electromotive force when it is inserted in a circuit, and he showed how to measure the internal resistance of a voltaic cell.

The measurement procedures described so far all involved the use of a galvanometer to show that the current was at a certain specific value, but it was never used to give a direct reading of the strength of the current. Wheatstone appreciated that 'it would greatly facilitate our quantitative investigations if we had a certain and ready means of ascertaining what degree of the galvanometric scale indicated half the intensity corresponding to any other given degree'. He described a switching arrangement which connected the galvanometer into a circuit either directly or through a resistance network. If the galvanometer had a resistance R then the network consisted of another resistance R in parallel with the galvanometer and a resistance $\frac{1}{2}R$ in series. The network plus galvanometer had the same resistance between its terminals as the galvanometer alone (so that when the network was introduced the external circuit was unaffected) but the current through the galvanometer was reduced to one half of its previous value (see Figure 9.4).

Wheatstone then showed how the circuit of a galvanometer could be changed so that it carried currents of $\frac{1}{2}$, $\frac{1}{3}$, $\frac{1}{4}$ etc. of some initial value. The corresponding deflections of the galvanometer could then be observed and the instrument calibrated for use as a direct reading instrument. He showed how to calculate the value of shunt resistance to be used with a galvanometer so that it could be used to measure currents of any desired magnitude.

The procedures described above for measuring resistance are not suitable for applications where small differences are involved. Wheatstone observed that Becquerel's differential galvanometer, which had two coils wound together but acting in opposition, would have been useful for such measurements, 'had it

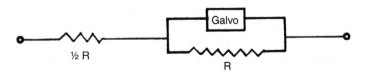

Figure 9.4   Circuit used in calibrating a galvanometer.

been an instrument as practically as it is theoretically perfect . . . But it is almost impossible so to arrange the two coils that currents of equal energy circulating through them shall produce equal deviations of the needle in opposite directions'. Mainly for that reason the differential galvanometer was not used in practice:

All the advantages, however, which were expected from this instrument [Becquerel's] may be obtained, without any of its accompanying defects, by means of the simple arrangement I am about to describe, which, moreover, has the advantage of being immediately applicable to any galvanometer, instead of requiring, as in the former case, the instrument to be peculiarly constructed . . . [Figure 9.1] represents a board on which are placed four copper wires, Zb, Za, Ca, Cb, the extremities of which are fixed to brass binding-screws. The binding-screws Z, C are for the purpose of receiving wires proceeding from the two poles of a rheomotor, and those marked a, b are for holding the ends of the wire of a galvanometer. By this arrangement a wire from each pole of the rheomotor proceeds to each end of the galvanometer wire, and if the four wires be of equal length and thickness, and of the same material, perfect equilibrium is established, so that a rheomotor, however powerful, will not produce the least deviation of the needle of the galvanometer from zero.

This circuit became known even in Wheatstone's own lifetime as the 'Wheatstone Balance', 'Wheatstone Bridge', or just the 'Wheatstone' (the *Oxford English Dictionary* gives 'Wheatstone' as an abbreviation for 'Wheatstone('s) bridge' from 1872), but Wheatstone did not claim to have devised the circuit. In a footnote he wrote:

Mr. Christie, in his 'Experimental determination of the Laws of Magnetoelectric Induction', printed in the Philosophical Transactions for 1833, has described a differential arrangement of which the principle is the same as that on which the instruments described in this section have been devised. To Mr. Christie must, therefore, be attributed the first idea of this useful and accurate method of measuring resistances.

The reference to Christie is in a footnote and there are references to the work of other people in the body of the text of Wheatstone's paper, which suggests that the references were an afterthought. We cannot tell whether the afterthought – if that is what is was – was his own, or whether it was prompted by a referee or other reader. Wheatstone ought to have known of Christie's work since it was in the *Philosophical Transactions*, but there was also an important difference between what Christie had done and what Wheatstone was doing. Wheatstone was using direct current, and finding a balance under steady state conditions. Christie was using pulses.[6] Christie's was a ballistic bridge, and the similarity in the circuits may not have struck Wheatstone immediately when he was devising his own arrangements.

Christie devised the circuit and method for a purpose similar to but not quite the same as Wheatstone's. His object was to compare the electromotive forces induced by magneto-electric induction in different metals. He knew that when two coils, one of iron wire and one of copper wire but otherwise identical, were connected to a galvanometer different currents would flow when a magnet was plunged into or withdrawn from the coils. The question he sought to answer was whether the difference in current was due entirely to the difference in resistance of the two metals, or whether a different electromotive force was induced in each metal. To answer the question he connected the two coils in series so that their electromotive forces were in opposition. He found that no current flowed and that therefore the same electromotive force was induced in each coil.

The credit is due to Christie for the important concept of comparing two electromotive forces by putting them – or a part of them – in opposition and making some adjustments until exact equality is indicated by a zero reading on a galvanometer. Wheatstone converted Christie's laboratory method into a practical procedure by making two advances. He showed how the method could be used for comparing resistances, and he put the essential circuit onto a board with terminals so that it became a piece of apparatus. He then publicized it.

The great advantage of the bridge method of measurement is that it is a 'null' method. Adjustments are made until the galvanometer indicates no current flow, so problems of calibration do not arise. Wheatstone did not specifically make this point in his paper, though he did remark that variations in the battery output did not matter.

After his Bakerian Lecture in 1843 Wheatstone does not appear to have made much further contribution to the theory or practice of electrical measurements. In 1845, however, Wheatstone and Cooke obtained a patent for various improvements in the ABC telegraph, and tucked away among the details of telegraph instruments there is a galvanometer with an iron core. Previous galvanometers all had their coils wound on a non-magnetic former. This is the earliest description of a galvanometer with a ferromagnetic core and it resulted in an instrument of greatly increased sensitivity.

In 1861 the British Association appointed a committee to consider the question of a standard of electrical resistance. The Electric Telegraph Company continued to use Wheatstone's own standard of resistance, though specifying one mile of the wire rather than one foot, after Wheatstone ceased to be connected with the company. Many other units were also employed and there was a widely felt need for an internationally agreed standard. The committee consisted originally of Professors A. Williamson, C. Wheatstone, W. Thomson, W.H. Miller, Dr A. Matthieson and Mr F. Jenkin. It issued its first report at the British Association meeting at Cambridge in October 1862, and several subsequent reports up to 1869. Wheatstone was a member of the committee throughout, but there is nothing in the reports to suggest that he took part in any of the experimental work. Some of the experiments were conducted at King's College by James Clerk Maxwell and others. Maxwell was Professor of Natural Philosophy at King's from 1860 to 1865.

In a footnote to his paper Wheatstone said that he had explained these instruments and processes to Professor Jacobi of St Petersburg at the beginning of August 1840. Jacobi had made a similar instrument, which Wheatstone said he called an 'Agometer'. This is mentioned, though not described in detail and not given a name, in Jacobi's paper 'On the Principles of Electro-Magnetical Machines' which he presented in person at the British Association meetings in Glasgow in September 1840. Jacobi says of his instrument:

It is destined to regulate the galvanic current, and is of value in many investigations of this kind. During my sojourn in London, Prof. Wheatstone has shown me an instrument, founded on exactly the same principles as mine, and with very inconsiderable modifications and differences. Now, it is quite impossible that he should have had the least notice of my instrument; but as it is probable that its use may be greatly extended, I must add, that while I have only used this instrument for regulating the force of the currents, he has founded upon it a new method of measuring these currents, and of determining the different elements or constants, which enter into the analytical expressions, and on which depends the action of any galvanic combination. It is principally to the measure of the electromotive force, by these means, that Mr. Wheatstone has directed his attention; and he has shown me, in his unpublished papers, very valuable results which he has obtained by this method.

Most of Wheatstone's paper on electrical measurements is still relevant. For everyday measurements reasonably accurate and stable direct-reading voltmeters and ammeters are now available. An ordinary voltmeter or ammeter today is only a galvanometer with a calibrated scale. In Wheatstone's time the magnetic materials available did not retain their magnetic properties without change over a period of time. Consequently it was not possible to have an instrument which would give a direct reading of current or voltage at any time without recalibration. In the last decade of the nineteenth century stable magnetic materials became available, making possible the manufacture of reliable voltmeters and ammeters. The circuits Wheatstone developed with resistances in parallel with and in series with an instrument, so that the instrument carries a specified fraction of the total current, are the basis of the modern multi-range testmeter. Rheostats and sets of coils of known resistance are still found in electrical laboratories. When the first edition of this book was written in 1974 it could be said that when extreme accuracy was required the modern engineer still used a bridge circuit. Now, however, with electronic digital meters which operate by counting tiny pulses, the bridge is rarely used but the Wheatstone Bridge was the principle circuit for precise electrical measurements for well over a century.

Notes

1   Arnold S. Lynch, lecture, 'The rise and fall of the AC bridge', Institution of Electrical Engineers' Wheatstone Measurements Lecture, 6 December 2000.
2   Jacobi's comments are in his paper 'On the Principles of Electro-Magnetical Machines' in the *Report* of the British Association for 1840.
3   G.S. Ohm, *Die Galvanische Kette Mathematisch Bearbeitet*, Berlin 1827. English translation, *Taylor's Scientific Memoirs*, **2**, 1841.
4   Fechner, *Massbestimmungen über die Galvanische Kette*, Leipzig, 1831.

5 Pouillet's memoir in which he described the sine and tangent galvanometers is in *Comptes Rendus,* **4**, 1837 and translated under the title 'Memoir on Volta's pile . . .' in Sturgeon's *Annals of Electricity*, **2, 1838.**

6 I am grateful to Dr Lynch (note 1 above) for bringing this point to my attention.

PART 3
THE TELEGRAPH

Chapter 10

Early telegraphy

The story of the electric telegraph as a practical proposition begins with the arrangements patented by Cooke and Wheatstone in 1837, but many inventors had tried, without great success, to make a telegraph before then.[1]

The earliest account of a workable electric telegraph was given in a letter signed only 'C.M.' in the *Scots Magazine* of 1753. This telegraph required 26 insulated wires running between the sender and the receiver. One wire corresponded to each letter of the alphabet. The operator sending a message had an electrostatic machine which he could connect to any one of the wires to indicate a letter. At the receiver the wires either ended with a spark gap or were arranged to attract small pieces of paper when charged. This telegraph had two fundamental disadvantages: it required a separate wire for each signal and because it depended on static electricity it needed a quality of insulation which could not be maintained over any great distance.

Sir Francis Ronalds built an experimental telegraph in the garden of his house at Hammersmith in about 1816. It also used static electricity but it needed only a single wire, which was sealed in a glass tube for insulation, then embedded in pitch in a wooden trough and buried in the ground. At both ends of the line a clockwork mechanism turned a dial with letters round it. The sender also had a frictional electric machine with which the wire was continuously charged. For receiving there was a simple electroscope consisting of two pith balls hanging from the wire on silk threads. Since they were similarly charged from the wire they hung apart. When the operator at the sending end wished to transmit a message he earthed the line at the instant the dial at his end indicated the desired letter. The person at the receiving end would see the pith balls fall together when the line was earthed, and note the letter indicated on his dial at that moment. The system was slow and it depended on the two dials keeping in step, but Ronalds showed that it would work over 150 metres of wire. He worked out a routine for synchronizing the two dials and wrote a pamphlet about his system. He thought of the possibility that mischievous individuals might damage the wire: his advice was 'hang them if you can catch them, damn them if you cannot, mend it immediately in both cases'.

A few years earlier S.T. von Soemmering in Munich had demonstrated a telegraph which used current electricity from a battery, rather than static electricity, but still required a separate wire for each letter or symbol. The receiver was a vessel containing a liquid such as acidulated water in which all the wires terminated. The liquid decomposed and liberated hydrogen when an electric current was passed through it. A message was sent by connecting the positive pole of the battery at the sending end to the wires corresponding to each letter in turn. Any other wire could be used for the return circuit to the negative pole of the battery. The person receiving the message noted from which wire a stream of hydrogen bubbles arose as each letter was transmitted.

In 1820 Örsted published his observation that a pivoted magnetic needle could be deflected by an electric current from a battery flowing in a nearby wire. It was soon found that the effect could be enhanced by arranging the wire in a coil surrounding the needle. The result was a device which could detect a small electric current very quickly, and this led to a number of experimental telegraphs in which one or more pivoted magnetic needles responded to the current in a wire. One of these was demonstrated by Professor Müncke at Heidelberg in 1836, and it was Müncke's demonstration that inspired William Fothergill Cooke to turn his attention to the telegraph.

Before the coming of the electric telegraph, news and messages could travel no faster than a horse or a carrier pigeon, except where the semaphore telegraph had been installed. Napoleon used the semaphore extensively to link his empire. The best known line of semaphores installed in Britain linked the Admiralty in London with Portsmouth, and had relay stations sending the messages on from one hill top to the next. This was an advance on the old beacons which could only signal a pre-arranged message. The government could send orders to the fleet by semaphore within a few minutes – provided it was daylight and not foggy. The authorities were well satisfied with the system, and when in August 1816 they were approached by Francis Ronalds with his ideas for an electric telegraph the Secretary of the Admiralty informed him bluntly that no telegraph other than the one then in use would be adopted.

As a boy of 15 Wheatstone saw Ronalds' demonstration telegraph. Perhaps the studious youth with a keen interest in everything scientific was a welcome visitor. There is nothing to suggest that Ronalds' experiment had any influence on Wheatstone when he turned to the electric telegraph. Ronalds used static electricity; Wheatstone made use of the magnetic effect of an electric current, discovered by Örsted four years after Ronalds' demonstration. Ronalds' telegraph depended on clockwork mechanisms keeping in step at each end, as did Cooke's; Wheatstone never used such a system.

The first customers for the new telegraph were the railways, which were only just being developed. Most of the main railway lines of Britain were built between 1830 and 1860. The railways and the telegraph were both linking centres of population, and since the railway companies had already purchased long strips of land the telegraph pioneers did not have to negotiate with a large number of separate landowners. The satisfactory operation of a railway system required rapid communications so that reports of train movements could be sent from one point to another. Even a simple telegraph could carry far more messages than were required for railway operation, and a service was soon established carrying messages for the public.

It is almost impossible to exaggerate the change brought about by electrical communication. The American colonies declared independence on 4 July 1776, but the news did not reach London until 21 August that year. The Battle of Trafalgar was fought on 21 October 1805, but the news of victory did not reach London until 2 November. In the 30 years up to 1868 – the year in which the telegraphs were taken over by the Post Office – the situation changed completely. In 1868 there were over 90,000 miles of telegraph wire in the United Kingdom, transmitting six million messages annually between 3,000 public telegraph stations. London had been in reliable and virtually instantaneous communication with Paris since 1852, with India since 1864 and with America since 1866. A large measure of the credit for this was due to Wheatstone.

We must now look separately at the work of both Cooke and Wheatstone before they met for the first time in 1837. This is particularly well documented because the two men quarrelled and in the course of arbitration proceedings both of them gave a detailed written account of their separate work on telegraphs. The arbitration papers and much of the correspondence between Cooke and Wheatstone are now in the archives of the Institution of Electrical Engineers.[2] There is an account of the Cooke and Wheatstone story in Geoffrey Hubbard, *Cooke and Wheatstone and the Invention of the Electric Telegraph*, 1965.

They first met at Wheatstone's house in Conduit Street on 27 February 1837. Both were interested in developing an electric telegraph, but their approaches were quite different: Wheatstone was pursuing a piece of scientific research, Cooke was embarking on a business venture. Cooke told Wheatstone that his intention was to take out a patent. Wheatstone told Cooke that his own intention was the advancement of scientific theory. When he had finished his experiments he would publish the results and 'allow any person to carry them into practical effect'. He had no plans to exploit the telegraph commercially himself.

Wheatstone's early telegraph ideas

Wheatstone had not published any plans for an electric telegraph, but a brief note in the *Magazine of Popular Science* for March 1837 reported some of Wheatstone's lecture demonstrations measuring the velocity of electricity and said that in a lecture given the previous June Wheatstone had explained how the apparatus could be converted into a telegraph. The note followed a short article about some electric telegraph experiments in Munich. The magazine was published after 25 February 1837 (it contained a list of patents granted that day). It has been suggested that the passage in question was added to the magazine after the type for the remainder had been set up and after Cooke's visit to Wheatstone, with the intention of proving that Wheatstone had considered making an electric telegraph before his first meeting with Cooke. Latimer Clark took this view in his *Memoir* of Sir William Fothergill Cooke where he wrote:

The information *given* could scarcely have come from anyone *but* Professor Wheatstone himself . . .; and it is hardly possible to doubt that, having been permitted to see the proof sheets of Gauss, Weber, and Steinheil's telegraphic experiments, and having, on the 27th of February, been visited by Mr. Cooke in reference to the same subject, he furnished the editor with the notes in question, which were inserted at the last moment in brackets.

There is no direct evidence that Wheatstone mentioned the idea of a telegraph in a lecture in June 1836, but a printed syllabus survives listing lectures given by him in May and June 1837 on 'The Measures of Sound, Light, Heat, and Magnetism and Electricity'. The last lecture, on 27 June, includes 'The Velocity of Electricity in a circuit of Copper Wire, four miles in length, measured by the revolving mirror'. It may reasonably be assumed that the same lecture was given in June 1836 and that since Wheatstone had long been interested in 'telegraphy' he should mention the possibility of developing an electric telegraph from his apparatus. If Wheatstone had arranged for the paragraph about his own work to appear in the *Magazine of Popular Science*, he would have made sure the report was correct and not one which could be disproved simply by asking his students!

When in 1840 the dispute which arose between Cooke and Wheatstone was submitted to arbitration, Wheatstone prepared a summary of his own telegraphic work before meeting Cooke. This summary formed the introduction to 'Professor Wheatstone's Case' submitted to the arbitrators:

The subject of telegraphic communication has for a long series of years occupied my thoughts. When I made in 1821 the discovery that sounds of all kinds might be transmitted perfectly and powerfully through solid wires and rods, and might be reproduced in distant places, I thought that I had an efficient and economical means of establishing a telegraphic (or rather a telephonic) communication between two distant places. Experiments on a larger scale, however, showed me, that though for short distances most perfect results were obtained, yet that the sounds could not be efficiently transmitted through very long lengths of wire. My ideas respecting establishing a communication of this kind between two distant towns are stated in my Memoir on the Transmission of Sound, published in the Journal of the Royal Institution for 1831. I desire also to refer to an extract from Ackerman's Repository for 1821, the year in which my experiments were first made known.

I afterwards turned my attention to the employment of electricity as the communicating agent. The suggestions and experiments of Cavallo, Reiser, Soemmering, Ronalds, Ampère and others had failed to produce any well-grounded belief in the practicability of such an application. This want of confidence resulted from the imperfect knowledge then possessed respecting the velocity and other properties of electricity; some philosophers made it travel only a few miles per second, others considered it to be infinite: if the former were true, the consequences to be deduced from it would not leave much room for hope; but if the velocity could be proved to be very great, there would be encouragement to proceed. I determined to investigate this point, and by means of some original experiments ascertained that electricity of high tension travelled through a copper wire with a rapidity not inferior to that of light through the planetary space, and I obtained abundant reasons for believing that electricity of all degrees of tension travelled with the same velocity in the same medium. The insulated circuit I established for these experiments enabled me to ascertain that magnetic needles might be deflected, water decomposed, induction sparks produced, &c., under properly arranged circumstances, through greater lengths of wire than had ever yet been experimented upon. In the year 1836 I repeated these experiments with several miles of insulated wire, and I ascertained (which had never been done before) many of the conditions necessary for the production of the various magnetic, chemical and mechanical effects in very long circuits. I also devised a variety of instruments by which telegraphic communication should be realized on these principles.

Wheatstone then referred to five 'contrivances which suggested themselves to me . . . of which some evidence besides my own exists'.

His first proposal was for a telegraph using electric sparks and the revolving mirror or a revolving prism. A 'particular contrivance for making and breaking the circuit' would cause a rapid sequence of sparks which would appear to the unaided eye as a single spark. When viewed by the revolving mirror the sparks would appear spread out and the sequence could be determined. It was not necessary to synchronize the revolving mirror. His 'rythmical telephone' (spelled thus; we read no more of it) was a modification in which 'the strokes of a bell are substituted for the sparks, and the voltaic current for the electric discharge'.

He had invented apparatus for making and breaking the circuit, 'which I have since employed in my new instruments, described in the third patent'. This was presumably the rotary switch used in the sending instrument of his first ABC telegraph, described below. It had been made by a Mr Kirby, 'a workman now in Mr Cooke's employ' in 1835 or 1836.

He proposed that the electrometer in Ronalds' telegraph should be replaced by 'a magnetic needle in a multiplier'. ('Multiplier' here refers to 'Schweigger's Multiplier', the use of a coil of wire carrying a current to deflect a magnetic needle in place of the simple straight wire used by Örsted when he found that a current in a wire could deflect a magnetic needle.)

Wheatstone's fourth point was that he had invented 'the commutating principle by which a few wires were converted into a number of circuits'. He had the keyboard which had been made before he met Cooke, and the workman who had made it could be called. The commutating principle will be seen to be important when their first patent is considered below. Finally, he had proposed using electromagnetic attraction to set off an alarm driven by an independent mechanism.

Wheatstone also claimed that he had discussed a possible telegraph on the London and Birmingham Railway with Mr Fox, engineer of the company, and that Mr Enderby had undertaken to prepare 'the insulating rope containing the wires' for trying 'an experiment across the Thames, from my lecture-room to the opposite shore'.

This according to Wheatstone, was the state of his work on the telegraph when he first met Cooke, and he could if necessary produce apparatus and witnesses to prove it. This appears to be a fair and accurate statement.

Cooke's early telegraph ideas

William Fothergill Cooke (1806–1879) was born at Ealing, Middlesex, where his father was a surgeon and a friend of Francis Ronalds. Later his father became Professor of Anatomy at the University of Durham. Cooke had a conventional classical education in Durham and then at Edinburgh University. He had no scientific education and the only mechanism with which he was acquainted when he began his telegraphic work was the musical box. He joined the East India Company's army at Madras and became an Ensign (a rank corresponding to the modern Second Lieutenant), but resigned because of ill health in 1833. While convalescent he travelled in Europe and started to make anatomical models in wax. He may have done this first as a teaching aid for his father; he found he was quite good at it, and could earn money in this way. However, something else was to catch his interest. In March 1836 Cooke saw an 'electro-telegraphic experiment' exhibited by Professor Müncke at Heidelberg. He wrote later:

I was so much struck with the wonderful power of electricity, and so strongly impressed with its applicability to the practical transmission of telegraphic intelligence, that from that very day I entirely abandoned my former pursuits, and devoted myself thenceforth with equal ardour, as all who know me can testify, to the practical realization of the Electric Telegraph; an object which has occupied my undivided energies ever since.

Müncke's apparatus had a voltaic battery using copper and zinc in acidulated water or wet sand, a piece of wire and a galvanometer comprising a card with a cross on one side and a straight line on the other fixed on a straw suspended from a silk thread and with a magnetic needle at the end suspended horizontally in a galvanometer coil. The apparatus was worked by moving the wires to make contacts. There were no switches or press buttons.

Within three weeks of seeing Müncke's experiment, Cooke had made 'my first Electric Telegraph of the galvanometer form'. It used six wires to form three circuits influencing the three needles so that 26 signals were possible. If each needle could move to the left or to the right or remain static then the three needles could be made to give $3 \times 3 \times 3 = 27$ different indications, including the case in which all the needles remained at rest. He had not then conceived the idea of using sequences of needle movements to give signals – each signal was indicated by a single combination of needle positions – nor did he appreciate that two or more circuits could share a common wire.

Cooke had devised three features which he considered essential to a practical telegraph system, and in his writings he laid great stress on the fact that his first telegraph incorporated these three features. The first was his Detector 'by means of which injuries to the wires, whether from water, fracture or contact, are readily traced'. This was a simple galvanometer. The most important feature in

Cooke's view was that his telegraph was reciprocal – messages could be sent both ways. Each terminus had both sending and receiving equipment in one instrument, and each circuit was provided with a 'crosspiece of metal' for completing the circuit at the extreme end when receiving signals. His third refinement was to install an alarm at each terminus, so that it did not have to be watched continuously, like the semaphore telegraph. His first alarm used an electromagnet to pull a sprung or weighted stop from a clockwork-driven bell. The idea of an alarm associated with telegraph instruments was certainly not new. (Ronalds proposed the use of a pistol, fired electrically, to call attention at the receiving station.) Only an unscientific man like Cooke would have thought the idea of a reciprocal telegraph very remarkable. The experimental telegraphs Cooke had seen were presumably demonstrated in one direction only, since that was adequate to show what could be done.

The alarm gave Cooke the idea for his 'mechanical telegraph' which he first thought of in March 1836. The arrangement was similar to Ronalds' telegraph in that only a single circuit was required which controlled the motion of an independently driven device. Cooke thought this was a much more promising line of investigation than the galvanometer type of telegraph. He returned to England in April 1836 and wrote a pamphlet in June and July of that year. This pamphlet was not published until 1857. In August 1836 he obtained an English patent for 'Winding up springs to produce continuous motion' but no specification was enrolled and no description survives. Presumably he found either that his invention was not novel or that it did not work. The fact that Cooke had had this patent is not mentioned in his publications.

Cooke obtained an introduction to the directors of the Liverpool and Manchester Railway, for he realized that the most likely people to use his telegraphs were the rapidly expanding railway companies. In January 1837 he showed the directors his pamphlet and instruments capable of giving 60 signals. The directors wanted a less complex telegraph with fewer signals, which Cooke quickly designed. He returned to London to have four of these made, and two were working at the end of April 1837.

Meanwhile Cooke experimented to find at what distance he could operate his 'mechanical telegraph'. He sought Faraday's help and Faraday visited Cooke at his lodgings in the Adelphi in November 1836. After the meeting Cooke wrote a letter to his mother, part of which is worth quoting for what it reveals of Cooke's understanding and intentions at the time:

I had good reason for believing that it was a well-ascertained fact that the galvanic fluid only imparted a magnetic quality to cold iron (or an electromagnet) when its course was short, and that the shorter the course the greater the attractive power. Hence electromagnets were made with several short thick wires for experiments requiring *quantity*, and one long thin wire where *intensity* was required! To set this point at rest I got an introduction (through Dr. Uwins) to Faraday, the 'King of Electro-Magneticians'. He received me yesterday, and I asked him to give me his opinion upon the instrument, and in the kindest manner he proposed calling this morning for half an hour, which he did, but stayed an hour and a quarter, entering with great interest into all the details. He finally gave me as his opinion that the 'principle was perfectly correct', and seemed to think the instrument capable, when well finished, of answering the intended purpose. He would not give an opinion as to the distance to which the fluid might be passed in sufficient quantity, but observed that if it be only for 12 or 20 miles it can be passed on again. He said in reply to my question: 'I am afraid of inducing you by my advice to expend any large sum in experimenting, but it would be well worth working out, and a beautiful thing to carry on in this manner a conversation from distant points; and the instrument appears perfectly adapted to its intended uses'.

Now I consider this highly satisfactory. He took his leave in a most friendly manner, but in a way which induces me to think he does not mean to take any further step in the affair. I asked his advice as to my way of proceeding in bringing it before the public when completed, but he declared his inability to advise me. I do not intend to expend anything more upon it myself, but hope to find someone who will take the risk in consideration of a fair remuneration in case of success. The difficulty of arranging the escapement I shall be able to overcome, beyond a doubt. You will see by what I have said that if I fail it will not be owing to any obvious error, which I ought to have detected myself; and that is a great relief to my mind, for I have felt the disgrace of failing from ignorance as to the principles of the power I employed, which would justly have exposed me to severe animadversion for presumption.

I showed him also my *perpetuum mobile* principle. He says it is original, to the best of his belief, but deemed it difficult to work out. He was so pressed for time that he could not enter into the nature of the means I proposed employing to attain my object.

Cooke's letter reads as though Faraday hastily excused himself as soon as he mentioned perpetual motion. He later said that Faraday had advised him 'to increase the number of the plates of the battery proportionately to the length of the wires'. This helped, but Faraday did not appreciate the significance of the resistance of the electromagnet coils. In February 1837 Cooke's friend and solicitor, Burton Lane, allowed him to arrange one mile of wire in his Lincoln's Inn Chambers. Arranging this length of wire occupied Cooke for three whole days, but the results were not satisfactory. Cooke reported these experiments in a letter to his mother, and added that he had consulted Faraday again and also Roget, the Secretary of the Royal Society, who referred him to Wheatstone. Cooke described Wheatstone as 'Professor of Chemistry at the London University'. The subject and the College were wrong, but he had the right man.

The partnership

Cooke and Wheatstone agreed in principle in March 1837 to form a partnership. Cooke first offered Wheatstone one-sixth of the profits in return for his scientific assistance but Wheatstone would only enter into partnership as an equal partner. The terms finally agreed were that Cooke should be business manager of the partnership and take 10 per cent of the profits as a manager's fee. The remaining profits would be divided equally between them and Cooke would have the sole right to act as contractor for installing telegraph lines. No partnership deed was signed at this time. During March and April they conducted many experiments together, and in May 1837 they applied for an English patent. The formal partnership agreement was set out in a deed dated 18 November 1837 after an informal arbitration of certain differences between Cooke and Wheatstone by Benjamin Hawes, MP. In a letter to his mother Cooke referred to this arbitration, though did not set out what the issues had been:

I went . . . with Mr. Wheatstone, and discussed with Mr. Hawes the heads of our agreement. We were occupied at it from 8 till 11 o'clock, when, everything being settled to our mutual satisfaction, our signatures were attached, and the papers placed in our lawyer's hands to embody their spirit in legal circumlocution. They contain all I wish, leaving me sole and entire manager in England, Scotland, and Ireland, with this one exception – that, before selling the patents or licenses, I am to obtain Mr. Wheatstone's acquiescence in the price. Nothing, of course, can be more reasonable.

Before 1852 England, Scotland and Ireland had separate, though similar, patent systems, and an inventor who sought protection for his invention throughout the United Kingdom had to make three separate applications. In accordance with English patent law at the time a petition for a patent of invention had to be presented first to the Home Secretary and it was then referred for a report to one of the Law Officers, usually the Solicitor General. If that report were favourable the petition went to the Sovereign in person for signature. Finally the papers went to the Lord Chancellor for the patent to be sealed with the Great Seal. The inventor had to take his petition from one office to the next and pay fees at each stage. Until the patent was sealed the invention had to be kept secret, for publication of the invention before the patent was sealed would invalidate the grant. The Law Officer usually required a written description of the invention (this preliminary description later became the provisional specification) and a full specification had to be enrolled in Chancery within a specified time, usually three or six months, after the patent was sealed.

Cooke and Wheatstone petitioned for a patent in May 1837. William IV signed the papers on 10 June after a delay caused by the King's illness. Cooke immediately wrote to his mother 'All is now safe'. It was not, of course, for the patent was not sealed until two days later. They were allowed six months to enrol the specification, and could include in the specification any developments during that six months which fell within the general description given to the Law Officer. The specification was enrolled on 12 December 1837. The original specification, which may be seen in the Public Record Office, is written on parchment strips 25 centimetres wide and about 66 long, sewn end to end. It occupies in all about 10 metres of a roll containing 12 specifications, all written in the same hand. The drawings, which are coloured, are folded and sewn into the roll.

Rival inventors

An interested person could oppose a patent application but it was difficult since the precise nature of the invention could not be disclosed until after the patent was sealed. The procedure was for the prospective opponent to lodge a 'caveat' with one of the Law Officers specifying the subject with which he was concerned and asking to be heard if a patent were sought relating to that subject. No patent specifications were printed officially until after the reform of the patent system in 1852, when all the English patent specifications from 1617 onwards were printed and also given numbers. Some journals published some specifications, but otherwise an interested person could only know the content of a specification by paying a fee to be allowed to see the enrolled copy. Scottish and Irish patents have never been published; the general procedure was the same as in England.

A caveat against patents for an electric telegraph had been lodged with the Solicitor General by Edward Davy, together with a description of his own telegraph experiments. His telegraph used individual wires for each letter and a common return wire. At first it was of the galvanometer type but he also described an arrangement in which an electromagnet lifted a cover to reveal a letter. Early in 1837 he conducted experiments in Regent's Park using a mile of copper wire. Then in March he heard rumours of Wheatstone's work – possibly he saw the note in the *Magazine of Popular Science*. Davy failed to persuade the Solicitor General to stop Wheatstone's and Cooke's patent, and he considered taking legal action to get it repealed, but did not do so because he expected to come to some agreement with them.

In October 1837 Cooke and Wheatstone began work on the specification and to help in this they engaged John Farey, 'the best consulting engineer in England' according to Cooke. Farey's signature is on the original drawings enrolled with the specification.

In August of the same year they applied for a Scottish patent and this was opposed by William Alexander of Edinburgh. Alexander had published a scheme for a telegraph earlier in the same year. In a letter to *The Times* on 8 July 1837 he described his telegraph which had a galvanometer needle and one wire for each letter and a common return wire. He had tried unsuccessfully to interest the Government. He failed to prevent Cooke and Wheatstone from getting a Scottish patent, although he may have reduced its scope. The opposition proceedings were described in a letter to Cooke's mother by Louisa Wheatley, who subsequently married Cooke:

I have the pleasure to tell you that the Scotch patent is now free from all opposition, and will be obtained immediately. William had an interview with Mr. Alexander on Thursday at the Lord Advocate's [of Scotland] office, and he agreed to accompany the Judge and Lord Lansdowne to see some experiments yesterday at Euston Square terminus. These proved so satisfactory that Alexander at once acknowledged the superiority of William's and Mr. Wheatstone's plans, and gave up his own. This was an agreeable way of arranging the matter, and William was pleased with Alexander's manner of yielding the point; though, of course, he saw he had no chance of succeeding.[3]

The Scottish patent was sealed on 12 December 1837 and the specification enrolled exactly six months later. The Irish patent was not sealed until 23 April 1838.[4]

It is of interest that Wheatstone, either alone or on behalf of himself and

Cooke, opposed other patent applications on at least three occasions. Edward Davy sought a patent in March 1838. The main feature of his patent was a recording telegraph in which the receiving instrument was arranged to pass a current through a moving paper tape soaked in a solution of a chemical such as potassium iodide, which decomposed when a current flowed leaving a coloured mark on the paper. Wheatstone opposed the application. The Solicitor General considered the case for several days, heard Wheatstone's arguments on at least two occasions and consulted Faraday before deciding that Davy should get a patent. It was granted on 4 July 1838.

In June 1838 Samuel Morse came to London and applied for a patent for his electric telegraph. The application was opposed by Wheatstone and Cooke and also by Davy on the ground that Morse had already published his invention in the *Mechanics' Magazine* of 10 February 1838. Morse seems not to have known that prior publication of an invention made it impossible to get a patent in England, and he regarded it as most unjust when the Attorney General, Sir John Campbell, refused Morse's application. Although Morse did not achieve the main object of his visit to London he found his meeting with English telegraph workers useful. He also found some consolation in the fact that while there he managed to obtain a seat in Westminster Abbey to see Queen Victoria's coronation on 28 June 1838. On his way to Paris the following month Morse wrote to his daughter:

Professor Wheatstone and Mr. Davy were my opponents. They have each very ingenious inventions of their own, particularly the former, who is a man of genius and one with whom I was personally much pleased. He has invented his, I believe, without knowing that I was engaged in an invention to produce a similar result; for, although he dates back into 1832, yet, as no publication of our thoughts was made by either, we are evidently independent of each other. My time has not been lost, however, for I have ascertained with certainty that the *Telegraph of a single circuit* and a *recording apparatus* is mine . . . I found also that both Mr. Wheatstone and Mr. Davy were endeavouring to simplify theirs by adding a recording apparatus and reducing theirs to a single circuit.

Later, in 1840, Morse considered a proposal by Wheatstone and Cooke that they should co-operate in America. Morse was tempted by the proposal, but eventually declined it.[5]

The Morse incident is well known. Another opposition, which has never before been published, is to be found in the Law Officers' papers in the Public Record Office. Most of the Law Officers left no records of their work in connection with patents, but Sir Thomas Wilde was one of the few who did. He was Solicitor General in April 1841 when Wheatstone lodged an objection to a patent application by Henry Pinkus. Pinkus obtained 11 patents between 1827 and 1843 for a variety of inventions. The earliest ones were concerned with gas lighting and most of the later ones with pneumatic railways. One patent, dated 24 September 1840, included a system in which train brakes could be operated by a motor, and the motor was energized by remote control with the aid of special conductor rails alongside the track and contacts on the train. Early in

1841 he submitted another patent application, and this was opposed by Wheatstone who probably thought that Pinkus might intend to adapt the previous invention so as to permit telegraphic communication with a moving train. The outcome of Wheatstone's action was that Pinkus gave an undertaking to the Solicitor General to the effect that it was no part of the object of his application to establish an electric telegraph. The patent was then granted on 14 May 1841.

The first telegraph patent

We now consider Cooke's and Wheatstone's patents. The patent granted to them and sealed on 12 June 1837 was the first English patent for an electric telegraph. The formal title of the invention was 'Improvements in giving signals and sounding alarums in distant places by means of electric currents transmitted through metallic circuits'. The official *Abridgments of Patent Specifications* summarizes the 'Improvements' in six points thus:

1. A five-needle telegraph. The symbols are indicated by the simultaneous deflection of two of the magnetic needles whose coils are included in the circuit, the required letter being found on a dial at the point to which they converge. In this signal apparatus the needles move in vertical planes, they being heavier at the lower ends to enable them to point vertically when not conveying signals; stops limit the angular motion of the needles, and they are astatic, one being within the coil, the other exterior to it on the face of the dial plate. To complete the circuit, a combination of finger keys is used, by which the line circuit is broken at the same time as the voltaic battery and signal apparatus are interposed; and by springs and studs, in connection with the finger keys, the current may be made to circulate in either direction. Five line wires are used in the above telegraph (one to each needle coil), or each needle may be deflected separately by having a sixth line wire for the return current. A four-needle telegraph on the same principle, with five line wires, is also described and shown.

2. Methods of insulating and supporting the line wires are described and shown, consisting of placing them in a resinous cement in channels in the rails of post and rail fences; or in various shaped tubes or troughs, the wires being previously covered with cotton, and varnished.

3. A method of deflecting telegraph magnetic needles by electromagnets. The current is made to magnetize two horseshoe electromagnets placed opposite to one another with the needle between their poles; similar poles are opposite, and deflection takes place according to the polarity (i.e. the direction of the electric current).

4. A method of 'sounding alarums in distant places'. The detent of a clockwork is removed by an electro-magnet, on the completion of the circuit, and is again interposed by a spring on the ceasing of the magnetism; or a hammer is made to strike a bell by similar means.

5. Sounding alarums in distant places by the aid of an additional 'voltaic battery'. A battery, devoted solely to working the alarum, and not belonging to the line circuit, has its local circuit completed by the electric current from the distant battery. For this purpose the line circuit deflects a needle on whose axis a forked wire is mounted, which completes the local circuit by dipping into mercury cups; or the local circuit

may be completed by the evolution of gas (caused by the line circuit), forcing mercury in a U tube into contact with a wire; the tube is closed at one end, and has a bulb containing diluted sulphuric acid into which the poles of the line wires are inserted. A method of connecting and disconnecting an alarum from the circuit in the needle telegraph, by a bolt connected to the alarum circuit raising a spring in the needle coil circuit, is described and shown.

6. Methods of ascertaining the precise place of injury to the line wires are described and shown; the coils of needles, mounted as in the needle telegraph, are included in the circuit between two of the line wires, one of which is defective, a station battery being connected to them; various points in the circuit are tried until one is found at which the needle is not affected; half the distance between this and the last effective point is then tried until the injury is found. A voltameter may also be used to test the completion of the circuit. To distinguish between the line wires, it is proposed to use varnish of various colours.

The first 'improvement' was essentially Wheatstone's invention, not Cooke's. The 'mechanical telegraph', on which Cooke was working when he first consulted Wheatstone, is not included in the specification at all. The official abridgment is misleading in one respect: it begins with a reference to 'A five-needle telegraph', but the principal invention was not limited to the use of *five* needles. It was the idea of a telegraph with several needles which indicated letters by the simultaneous deflection of two needles. Five was the preferred number, according to the patent, but the first complete telegraph equipment demonstrated by Cooke and Wheatstone to a prospective customer was a four-needle arrangement which could indicate only 12 letters. This instrument does not survive and there is no record of which 12 letters were selected. It was demonstrated between Euston and Camden Town, on the railway then being constructed between London and Birmingham. The plan was for trains leaving London to be rope-hauled up the hill to Camden Town where a locomotive would be coupled on. Trains returning to London were detached from the locomotive at Camden and ran under gravity into Euston. Robert Stephenson and the directors of the company required a means of signalling between Euston and Camden Town so that the person operating the winding engine would know when the train was ready to start. Cooke obtained an introduction to Stephenson and arranged a demonstration. Stephenson was sufficiently impressed to provide men and materials for Cooke to arrange a larger demonstration for the directors. They were impressed, but did not adopt an electric telegraph. Instead they employed a simple pneumatic arrangement which blew a whistle in the engine house when the train was ready.

The five-needle telegraph (see Figure 10.1) was adopted for the first commercial electric telegraph line which Cooke, as business manager of the partnership, installed in 1838 along the Great Western Railway between Paddington and West Drayton. *Six* wires were used so that any needle could be operated singly to give additional signals.

The two main features which distinguish Cooke's and Wheatstone's four- and five-needle telegraphs from all earlier telegraphs, including those made

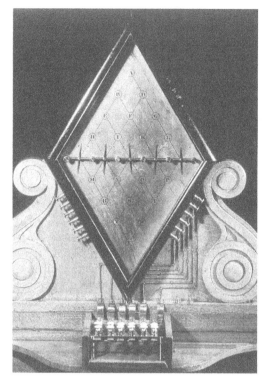

Figure 10.1   The five-needle telegraph.

by Cooke, are the use of the 'permutating principle' and the use of weighted vertical needles.

The 'permutating principle' was the switching arrangement whereby the telegraph wires were connected in various ways to form different circuits for the different signals (Figures 10.2 to 10.4). When in 1840 Cooke was preparing his case for the Arbitrators in his dispute with Wheatstone, he included a drawing of 'Prof. Wheatstone's Permutating Key-board and only Instrument at the date of the Partnership'. Since Cooke, when trying to minimize Wheatstone's contribution, freely attributed the permutating principle to Wheatstone, there can be little doubt that it was indeed Wheatstone's invention.

After the Paddington to West Drayton telegraph, no more telegraph systems were built using the permutating principle. The attraction of the five-needle telegraph was its extreme simplicity of operation. With a few minutes' instruction anyone could transmit a message on it, and anyone who could read could receive a message without prior training. However, the need for so many wires made the five-needle telegraph extremely expensive to instal, and when the line was extended to Slough in 1843 Cooke installed two-needle instruments using coded signals. Before long almost all telegraphs were single-needle instruments operated through a single wire with earth return. Cooke installed a telegraph

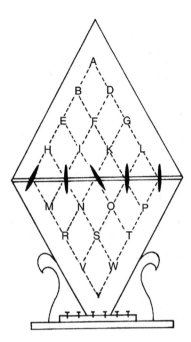

Figure 10.2   *The five-needle telegraph showing the arrangement of letters on the dial. To indicate a letter two needles are deflected, one in each direction. As shown the letter 'E' is being indicated.*

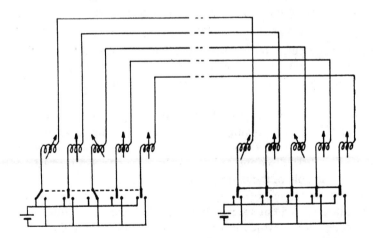

Figure 10.3   *The circuit of a pair of five-needle telegraphs. As shown the keys of the instrument on the left are pressed to indicate the letter 'E', which will be shown of the dials of both instruments.*

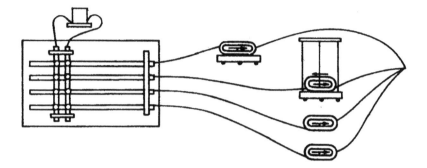

Figure 10.4   Cooke's drawing of Wheatstone's permutating keyboard in use.

system on the London and Blackwall Railway in 1840 which used five single-needle instruments. However, this telegraph was not designed for sending ordinary messages but merely to transmit a signal meaning 'stop' or 'ready' from each of five stations. A letter Wheatstone wrote to Cooke in January 1838 shows that at that time the permutating principle seemed important to the partners, perhaps because they needed to convince prospective users that their telegraphs not only worked but were simple to use!

> Previous to our patent no person had ever proposed otherwise than to employ a complete circuit (i.e. two wires) for each magnetic needle, the most important original feature in my instrument was that the same wire should be capable of forming different circuits according as it is conjoined with other wires. After the patent was sealed a notice of some of my experiments appeared in the Scotsman; some weeks subsequently there appeared in the same paper an account of what Mr. Alexander intended to do, and after a long interval a description of a model which he had produced. There is no doubt that in our Scotch Patent we must limit the application of the permutating principle.

The passage quoted above should probably have ended '. . . we must limit the application *to* the permutating principle'. The Scottish patent was not in fact restricted to the permutating principle, but it did have some additional description and claims relating to matters which were included in the second English specification.

The Science Museum has two of Wheatstone's permutating keyboards. The five pairs of buttons are arranged so that when the buttons are not pressed the five line wires are connected to a common contact bar. On pressing a button the line wire is first disconnected from the common contact bar and then connected to one pole of the battery (Figure 10.5).

The other main feature of the five-needle telegraph was the use of weighted vertical needles. The patent specification included a specific claim to a telegraph using magnetic needles which move in a vertical plane between end stops and have one end heavier than the other. This feature was equally applicable to single-needle telegraphs and it gave the patent a continuing importance when the five-needle system had been abandoned. It seems to have been Wheatstone's

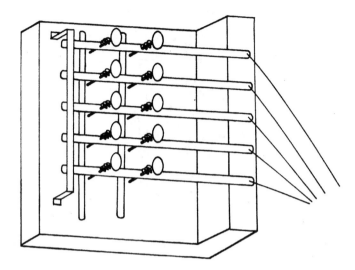

Figure 10.5   Wheatstone's permutating keyboard.

invention to use a vertical needle, for Cooke said later 'immediately after the formation of our partnership, Professor Wheatstone invented his Diagram with vertical astatic needles, a most important practical improvement'. Wheatstone probably also introduced the idea of weighting the needle so that it returned to its initial position after the current in the coil ceased. In early galvanometers the needle was restored to its initial position either by a spring or by the earth's magnetism. Cooke's drawings both of his own instruments and Wheatstone's instruments before their first meeting include galvanometers with a torsion suspension, and his drawing of Wheatstone's instruments also shows an astatic galvanometer.

The five-needle telegraphs used between Paddington and West Drayton had astatic needles, each pair operated by a single coil. The patent included an alternative arrangement (see Figure 10.6) with two horseshoe electromagnets deflecting the magnetic needle. This more elaborate arrangement does not appear to have been used in practice, though it would have ensured that the needle turned to a definite position without the use of end stops.

The patent specification suggested laying the telegraph wires in slots in a strip of wood. This was used between Euston and Camden Town, but not later. It was probably Cooke's idea, and it is of interest that the specification showed five wires (Figure 10.7). The Paddington to West Drayton telegraph had a sixth wire and a sixth pair of keys on each instrument so that single-needle signals could also be sent. According to Wheatstone the additional wire and keys were Cooke's idea, but Wheatstone did not like the modification because 'the simplicity and symmetry of my arrangement was destroyed, while no advantage whatever was obtained'.

The patent specification included claims for 'sounding alarums in distant

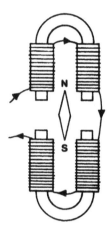

Figure 10.6   An alternative arrangement of the needles and electromagnets, shown in the 1837 patent.

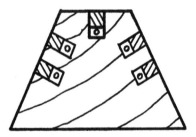

Figure 10.7   Method of laying telegraph wires in grooves in a strip of wood.

places' by 'the attractive force of occasional or temporary magnetism' in any of three ways. The electromagnet could itself operate the bell, or it could withdraw a detent from a clockwork mechanism which then operated the bell, or it could be used to complete another circuit so that a local battery operated the bell. For the last mentioned possibility two relays were described, though the word 'relay' was not used (see Figure 10.8). Professor Henry recorded that Wheatstone showed him a relay when he was at King's College in April 1837.

This review of the content of Cooke's and Wheatstone's first patent specification would not be complete without mentioning two matters which are conspicuous by their absence.

First, the specification makes no mention of the telegraphs on which Cooke was working when he first went to Wheatstone. Writing in 1856 Wheatstone recalled his early acquaintance with Cooke and remarked:

On no occasion during Mr. Wheatstone's acquaintance with Mr. Cooke and his 'practical realities' was Mr. Cooke's instrument exhibited to him in action, even in a short circuit; it was, after it had been proposed to be inserted in their first patent, omitted as useless, and

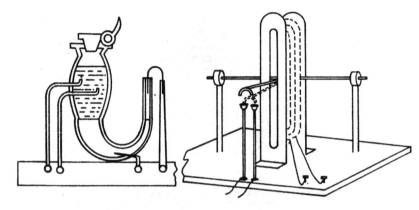

Figure 10.8   Two designs of relay shown in the 1837 patent.

Mr. Cooke, when he took out the second patent himself, did not think it of sufficient importance to mention it there.

Cooke responded that the instrument had been omitted from the first specification for lack of time in drafting, but he did not say why he omitted it on the second occasion.

The second matter is Ohm's Law, and the fact that Wheatstone alone had the necessary understanding of electric circuit theory to design a telegraph system which would work through several miles of wire. It is not surprising that Cooke could not do it. It may be surprising that Faraday and Roget could not do it. Wheatstone did it. In his case to the arbitrators he set out what he considered to be his most important contribution to the first practical telegraph in the following words:

The most important point of all was my application of the theory of Ohm to telegraphic circuits, which enabled me to ascertain the best proportions between the length, thickness, &c. of the multiplying coils and the other resistances in the circuit, and to determine the number and size of the elements of the battery to produce the maximum effect. With this law and its applications no persons who had before occupied themselves with experiments relating to Electric Telegraphs had been acquainted.

Among Wheatstone's papers at King's College are two sheets with calculations on telegraph circuits. They are probably lecture notes and they show the effect on the receiving instrument of various changes in the circuit of a galvanometer-type telegraph:

In the electro-magnetic telegraph, the length of the wire being very considerable the resistance in it is much greater than that in other parts of the circuit. If we call the electromagnetic force 10, the resistance of the battery 2, and the resistance corresponding to the length of the wire 1000, we may illustrate the following conclusion. ['A' apparently represents the current.]

1.   The deviation of the needle will increase rapidly with the number of pairs [of battery plates]. With one pair $A = 10/(1000 + 2) = 10/1002$. With ten pairs $A = 100/$

Figure 10.9   A Cooke and Wheatstone two-needle telegraph. The five-needle telegraph was too complicated and the necessary wires too expensive. The railways quickly adopted simpler telegraphs using either one or two needles only.

(1000 + 20) = 100/1020. With 100 pairs A = 1000/(1000 + 200) = 1000/1200). With 1000 pairs A = 10000/(1000 + 2000) = 10000/3000. The limit of increase will be E/R = 10/2.

2.   Increasing the size of the plates will not materially increase the effect. With one pair having double surface A = 10/1000 + 1 = 10/1001. With ten pairs A = 100/1010. With

100 pairs A = 1000/1100. With 1000 pairs A = 10000/2000. Comparing these with the preceding numbers it will be seen that the advantage is not considerable, but the advantage of using larger plates becomes more appreciable as the number of plates is increased.

3.   If the diameter of the wire is increased while the size of the plates and the length of the wire remains the same, the proportionate advantages will be more felt as the number of plates is less. With one plate *A = 10/502. With 10 plates A = 100/520.* With 100 pairs A = 1000/700. With 1000 pairs A = 10000/2500.

4.   If the size of the plates and the diameter of the wire be simultaneously increased the force of the current will be increased in the same proportion. With 1 pair A = 10/ (1000 + 2) the size of the plate and the diameter of the wire being doubled A = 10/ 501; the force of the current is therefore double. With 100 pairs *A = 1000/1200*; the size of the plates and the diameter of the wire being doubled *A = 1000/600.*

5.   If the number of pairs, and the length of the wire be increased in the same proportion, the force of the current will remain the same; for then the total resistance will increase in the same proportion as the electro-motive force.

Here we see Wheatstone at work, applying Ohm's Law to the practical problems of the telegraph.

Notes

1   A good account of the early electric telegraph experiments is given in J.J. Fahie, *History of the Electric Telegraph to the Year 1837*, London, 1884. There are numerous recent histories, including J.L. Kieve, *Electric Telegraph – a Social and Economic History*, 1973, and Ken Beauchamp, *History of Telegraphy*, 2001.

2   The principal sources of information about Cooke and Wheatstone and their early telegraphs are Cooke's letters and the publications arising from their dispute. Cooke's letters are in the Archives of the Institution of Electrical Engineers. Extracts have been published in F.H. Webb (ed.), *Extracts from the private letters of the late Sir William Fothergill Cooke, 1836–39, relating to the Invention and Development of the Electric Telegraph*, London, 1895. All the letters quoted here are included in the *Extracts*, except his correspondence with Wheatstone.

Nothing was published at the time of the arbitration in 1840. Cooke published *The Electric Telegraph: was it invented by Professor Wheatstone?* in December 1854. Wheatstone published *Answer to Mr Cooke's Pamphlet* in January 1856, and Cooke then published his *Reply* in March 1856. The three pamphlets were republished and the arbitration papers published for the first time by Cooke in 1856 and 1857 in two volumes also called *The Electric Telegraph: was it invented by Professor Wheatstone?*

There is an account of the Cooke and Wheatstone story in Geoffrey Hubbard, *Cooke and Wheatstone and the Invention of the Electric Telegraph*, 1865.

3   Louisa Wheatley's letter is quoted in Fahie, *op. cit.*

4 The specification of the Scottish Patent is unpublished, and may be seen in the Scottish Record Office in Edinburgh.
5 Edward Lind Morse, *Samuel F. B. Morse; His Letters and Journals*, Boston and New York, 1914.

Chapter 11

The practical electric telegraph

The demonstrations at Euston in 1837 did not lead to an electric telegraph installation and the five-needle system used between Paddington and West Drayton which opened in July 1839 was not adopted anywhere else, but the partners obtained considerable experience in the practical problems of installing and maintaining a telegraph system. In 1838 Cooke obtained a patent in his own name with the same title as their joint patent of the previous year. It was concerned mainly with very practical matters and had six features which Cooke said he was listing with letters rather than numbers in order to avoid confusion with the numbered claims of the first patent. The features were:

(a) connecting a telegraph instrument at an intermediate point in the line;
(b) having an alarm on the intermediate apparatus which could be sounded from one end station while the intermediate apparatus was receiving signals from the other end station;
(c) temporarily connecting a portable apparatus at intermediate points where special junction boxes were provided;
(d) protecting the wires by drawing them into iron pipes;
(e) using magnetic attraction rather than gravity (or as well as gravity) to restore the needles to their initial position;
(f) an improved alarm using an electrically released clockwork mechanism but no local battery.

Isambard Kingdom Brunel was the engineer of the Great Western Railway and he was responsible for the contract between the Railway and Cooke and Wheatstone. He decided that the idea of drawing the telegraph wires into metal pipes, rather than burying them in grooves in a strip of wood, should be adopted for the telegraph between Paddington and West Drayton. Cooke organized experiments to ensure that the workmen could do it satisfactorily. In one of his letters to his mother he described the experiments: 'I mean to lay down about a quarter of a mile in the garden over and over again, till each man knows his duty. The materials will do again on the main line, except the wires, which will have their jackets worn out; and even they can be burnt, cleaned, and re-covered'. The system was effective but expensive, and when the telegraph was extended to Slough the wires were suspended from insulators on poles. The system of wires drawn into pipes was used by Cooke on one other installation, the London and Blackwall railway. In that case the pipes were filled with compressed air to keep out moisture – a remarkable anticipation of the modern gas-filled power cable.

The improved alarm removed the detent from the clockwork mechanism by arranging for the magnetic needle to strike it, rather than just to press on it. This avoided the need for a local battery and proved very reliable.

It is not at all clear how much Wheatstone was personally involved in the installation of the telegraph along the Great Western Railway. The partnership agreement gave Cooke the right to act as contractor for installing telegraphs in this country, but Wheatstone was actively engaged in exploiting their inventions in Belgium and France, countries he visited several times at this period. Wheat-stone was on close terms with Brunel and was one of a very select group of people who accompanied Brunel on a speed trial. Another was the painter John Martin, a man fascinated by all engineering matters, whose son Leopold has left a dramatic account of the affair in his *Reminiscences*:

On the completion of the Great Western Railway to Bristol, in the year 1841, my father received a polite invitation from Mr. I. K. Brunel, its engineer, and son of Sir Isambard Brunel [usually known as Sir Marc (Isambard) Brunel], of Thames Tunnel renown, to accompany him on an experimental trip, in which it was his full intention to test the power and capabilities of a broad gauge engine as to the speed to be attained. As yet no experiments had been made by authority. The arrangements on this present occasion for a positive test were most complete. During a certain period of the morning, the line, for a given distance, was to be kept clear and free from traffic. A most experienced engine-driver was to take charge of the engine, one of the most powerful the company had constructed. On the morning appointed, my father's friend, Mr. Wheatstone, accom-panied him to meet Mr. Brunel at the Paddington Station. Mr. I.K. Brunel was also attended by time-keeping clerks and a very old office assistant. The party at once drove on the engine – there were no carriages – to the Southall Station, at which point the experiments were to commence. The first object was to try and discover the time in which full speed could be obtained, and in how short a distance the engine could be pulled up or brought to a stand. But the chief object Mr. Brunel had in view was to ascertain the highest rate of speed. This, in fact, was the only point of real interest.

The start took place from Southall, the engine being provided with a thick plate glass screen in front to protect those upon it from the strong current of air. With Mr. Brunel's eye on the steam pressure gauge and hand on the safety valve, we were off, and in half a mile we were running at 'top speed', the time-keepers busily at work. To the great satisfac-tion of Mr. Brunel and the astonishment of all, it was discovered that the distance of nine miles from the station at Slough had been run in six minutes, or at the rate of ninety miles an hour – a very different result from that which Mr. Stephenson's early calculations would have led one to expect.

'ABC' telegraphs

In January 1840 Cooke and Wheatstone obtained another joint patent (see Figure 11.1). This was concerned with their first 'ABC' telegraphs, so called because the receiving instruments indicated actual letters of the alphabet instead of giving coded signals by deflecting one or more needles. The ABC telegraphs were partly Wheatstone's invention and partly Cooke's. Thanks to Cooke's desire to set everything out in great detail in the arbitration papers it is possible to say which parts are due to each man.

The principle of the ABC telegraphs or 'mechanical telegraphs', as Cooke

Figure 11.1   *Wheatstone's ABC telegraph transmitter of about 1840. The person sending a message turns the wheel, always in the same direction, and stops as each letter of the message reaches the index. A pulse is transmitted as each letter passes the index.*

called them, was that letters or other symbols arranged around a circle in the receiver were indicated successively by a rotating mechanism controlled by the sending instrument (or 'communicator') (see Figure 11.2). To indicate a letter the motion of the receiver was stopped briefly at the appropriate point. Two main types of mechanism were described in the patent specification: one with an independent local power source, such as a clockwork motor, and one without. The receiving instrument (termed the 'signal apparatus' in the specification) received from the sending instrument a number of electric pulses corresponding to the number of letters which had to be passed before the receiver indicated the next required letter. In receivers with a local power source the drive was controlled by an escapement whose pallets were rocked either by two electromagnets energized alternately or by one electromagnet and a spring. If there were no local power source then the mechanism was driven through a pawl and ratchet arrangement, again actuated either by two electromagnets energized alternately or by a single electromagnet and a spring. The communicator had to send the appropriate pulses in either one of two circuits, and this was done either by switching a battery-powered circuit or by a magneto-electric machine which gave a pulsating output.

Cooke had made several 'mechanical telegraphs' before his first meeting with Wheatstone, and it was his inability to operate these through a long length of wire which led him to consult Wheatstone. The 1840 patent included some developments of these early mechanical telegraphs of Cooke, and it is therefore appropriate to give some consideration to them. Cooke has left drawings showing their mechanism but no written explanation of their operation. The description given here is the present writer's interpretation of the drawings.

Cooke's first mechanical telegraph was based on a musical box mechanism. It had a clockwork-driven rotating drum which could be started and then stopped at any point by energizing and then de-energizing an electromagnet which pulled against a spring to withdraw a detent from the mechanism. The idea was that there would be similar mechanisms at each end of the telegraph line and that these would run at exactly the same speed. The person sending a message would start the mechanisms by energizing the circuit of the electromagnets, and then stop when the desired letter was indicated on the instrument at his end and therefore also on the instrument at the far end, which the person receiving the message would be watching. This telegraph was made in March 1836. There was no means for ensuring that the drums kept in step, and the switching was performed by dipping a wire into mercury.

Cooke's second instrument, which he dated August 1836, was much more advanced. The receiver had an escapement controlled by a pendulum which could be held at either end of its travel by one of two electromagnets connected in series. The communicator would have been controlled by a pendulum of the same length to keep the two instruments in step, though this is not shown in Cooke's drawings. The pendulum of the communicator closed a pair of contacts at each end of its swing. If the communicator was stopped when a contact was made, then the receiver would also stop because its pendulum would be held by

Figure 11.2 Wheatstone ABC telegraph receiver of about 1840, with its case removed. Letters of the alphabet are indicated by a needle which moves forward one letter for each pulse received. The person receiving the message notes the letters at which the needle stops.

one of the electromagnets. In this way the dial or pointer of the receiver would indicate all the letters successively until the desired one was reached, when the sender would stop his instrument. Although Cooke dated this instrument to August 1836 when he was giving an account of his work, it is probably the instrument he was awaiting throughout October and November of that year. Cooke's letters at that time tell of his problems: 'My Clockmaker has again disappointed me . . . the balance work has been broken . . .', 'My instrument was to have been finished this morning, but . . . a wheel in the escapement was wrong', '. . . my instrument – which *should* make its appearance at six o'clock . . .'. Eventually the news is 'My instrument came home late last night, but does not answer. I have, however, arranged another plan for keeping the time, which I shall try.'

It seems from the last quotation that Cooke found insuperable problems in 'keeping the time' (that is, keeping communicator and receiver in synchronism). The drawing of his third telegraph, dated February 1837, shows it to be more like his earlier and cruder apparatus of March 1836 than the August one. It again has what appears to be a free-running 'musical box' type of mechanism which is stopped by a sprung detent until an electromagnet is energized. It has six possible stopping positions (and could therefore give six different signals), whereas Cooke's first instrument has 15. However, the design of the communicator is much improved. It has a drum with six projections representing the six stopping positions. There are six keys, one for each signal, and when a key is pressed the communicator mechanism is released and runs until the corresponding projection is reached. While the communicator is running the circuit will be completed and the detent withdrawn from the receiver, so that the receiver runs also. Provided that the receiver keeps in time with the communicator it will indicate the desired signal when the communicator stops running and breaks the circuit.

The two ABC telegraphs described in the specification of the 1840 patent differ fundamentally from Cooke's 'mechanical telegraphs' described above in that they do not depend on two independently running mechanisms. One telegraph was designed by each partner. Cooke's signal instrument, or receiver, was developed from his receiver of August 1836, but whereas the earlier instrument had a pendulum which could be held at either end of its travel by one of two electromagnets connected together, his 1840 one had an escapement-like mechanism which was driven by two independently operated electromagnets. The sending instrument was a simple two-way switch and the person sending the message had a receiving instrument also so that he could see the signal he had sent. These instruments were never used in practice, nor was any account of them ever published.

Wheatstone's first ABC telegraph was referred to by Cooke as 'Wheatstone's Escapement Telegraph' because, in one version at least, it employed an electromagnetically operated escapement similar to Cooke's. Whereas Cooke's had two electromagnets and three wires, Wheatstone preferred a single electromagnet (and two wires) to move the escapement in one direction and a light spring to move it in the other. A clockwork mechanism advanced the dial by one letter at each movement of the escapement. In another version Wheatstone used a pawl and ratchet arrangement to move a wheel forward by one tooth each time the single electromagnet was energized. Several instruments of the ratchet type survive. Some in the Science Museum are mounted in neat wooden cases which also incorporate an electric bell. Wheatstone's name alone appears on them.

The communicator or sending instrument of Wheatstone's ABC Telegraph was much more advanced than the two-way switch used by Cooke. In the specification it was described as having 'a metallic circle inlaid with wood'. The communicator had a dial with letters around it corresponding to the letters on the receiver, and radial arms with which it could be turned. A contact spring pressed on the edge of the 'metallic circle', and as the dial was turned the 'metallic circle' turned also. The spring pressed alternately on metal and wood and so the circuit was alternately made and broken.

The last topic considered in the 1840 patent was alarms. Cooke's 1838 patent had included a clockwork-powered alarm in which a detent was removed from the clockwork mechanism by being hit by a moving magnetic needle. By 1840 Wheatstone had devised an even more sensitive arrangement, as Cooke grudgingly admitted:

The simple alarm described in the second specification [Cooke's of 1838] is now in daily use upon the Blackwall and Great Western Railways, and works with unfailing certainty. I may in passing observe, that should we hereafter find the power of the needle insufficient for its object at very great distances, Professor Wheatstone's further and later improvement, described in our Third English Specification [1840], may prove a valuable invention. Instead of removing the detent of the alarm by the blow of the needle itself, Professor Wheatstone interposes the needle between a hammer moved by clockwork and the detent, and thereby transmits the blow of the hammer to the detent; which the hammer cannot reach when the needle is not interposed.

Dispute and arbitration

Several references have been made to the dispute between Cooke and Wheatstone which led to a formal arbitration in 1840–1841. The quarrel arose because Cooke alleged that Wheatstone was claiming an undue share of the credit for the invention of the telegraph. By an agreement dated 16 November 1840 Cooke and Wheatstone agreed 'that the relative positions of the said parties should be ascertained by arbitration'. Nothing was published concerning the dispute until December 1854, when, after a renewal of the controversy, Cooke published a pamphlet entitled *The Electric Telegraph: Was it invented by Professor Wheatstone?* In the opening pages Cooke made the point that their patent of June 1837 was the first patent anywhere in the world for a practical electric telegraph, and he continued:

It has been supposed by many persons that this invention of the Electric Telegraph, in the year 1837, was the result of a lengthened process of investigation and experiment on the part of the eminent Professor named in my title page, aided towards the end of his labours by a partner, variously misrepresented as the capitalist, the mechanic, or the man of business, with whom he is understood to have associated himself just before taking out his patent. Documents are now in the press which will tell a very different tale; documents which were already in print thirteen years ago, and of which a copy has remained in Professor Wheatstone's possession ever since. They will convince the most prejudiced, that it was not from his philosophical information, nor from his experimental ingenuity, that Professor Wheatstone acquired his position as one of the patentees of the first practical Electric Telegraph; but from a communication made to him in confidence by the writer, who was then completing the practical invention, and was about to take out a patent for it; who was in possession of practical Electric Telegraphs already made by him, and fit for practical use . . . ; and who, having consulted Professor Wheatstone (before the Professor had done anything practical at all) as a scientific man, on a scientific question affecting the proportions of part of the apparatus, and having tried experiments with him upon the point submitted for his advice, was induced, by the adviser's scientific acquirements, and by pecuniary considerations, to admit him to a share in the patent, as a second partner . . .

. . . The invention at once became a subject of public interest; and I found that Mr. Wheatstone was talking about it everywhere in the first person singular . . . At length, in 1840, I required that our positions, relatively to the invention and to each other, should be ascertained by arbitration.

The 'documents now in the press' were the statements submitted to the arbitrators and the arbitrators' award. Cooke published the documents and a pamphlet reiterating his complaints against Wheatstone after seeing an anonymous article in the *Quarterly Review* for June 1854 which summarized the history of the telegraph mentioning Wheatstone but not Cooke. Wheatstone published an *Answer* in January 1846 and Cooke's *Reply* appeared in March 1846. The *Answer* seems to have been Wheatstone's only publication on the subject. There were further publications arguing Cooke's claims from 1866 onwards when Wheatstone's knighthood was under consideration. Wheatstone was knighted in 1868 and Cooke in 1869.

The arbitrators in 1841 were Sir Marc Brunel, nominated by Cooke, and Professor Daniell, nominated by Wheatstone. Their task was to determine 'in what shares, and with what priorities and relative degrees of merit, the said parties hereto are co-inventors of the Electric Telegraph'. According to the statements submitted to the arbitrators, both partners were prepared to call witnesses but there is no record of any witnesses being heard. When the proceedings began it was found that Cooke had had printed 1,000 copies of a volume containing each partner's statement and an address by Cooke's solicitor in response to Wheatstone's statement.

Wheatstone objected to the preparation of such a volume, intended to be published 'without the fair accompaniment of Mr. Wheatstone's reply to Mr. Cooke's case'. Cooke defended his action by saying that the other proceedings could be published later. According to Wheatstone, the address by Cooke's solicitor was 'condemned for its form and spirit so strongly by the arbitrators, that they refused to proceed unless it was withdrawn, and the printed papers placed at their disposal'. The arbitrators took possession of the 1,000 volumes and these were destroyed.

The arbitration papers and related documents are of interest as the principal source of information about Cooke's and Wheatstone's work at the time. The arbitration did not really settle anything, but the arbitrators produced a 'statement of award' which satisfied both partners:

Whilst Mr. Cooke is entitled to stand alone, as the gentleman to whom this country is indebted for having practically introduced and carried out the Electric Telegraph as a useful undertaking, promising to be a work of national importance; and Professor Wheatstone is acknowledged as the scientific man, whose profound and successful researches have already prepared the public to receive it as a project capable of practical application; it is to the united labours of two gentlemen so well qualified for mutual assistance, that we must attribute the rapid progress which this important invention has made during the five years since they have been associated.

M^{c}I^d BRUNEL
J.F. DANIELL
London, 27th April, 1841

Both Cooke and Wheatstone stated that they were satisfied with the award. The arbitration proceedings were conducted in private and the arbitrators clearly wanted the proceedings and their statement of award to remain unpublished. By the agreement setting up the arbitration both parties were at liberty to publish if they wished and Cooke did so 13 years later, but the statement of award must be read bearing in mind that all concerned intended and expected it to be kept private. Cooke was well satisfied with the statement that he had 'practically introduced and carried out the Electric Telegraph as a useful undertaking', and he was quite happy for Wheatstone to be 'acknowledged as the scientific man'. It may be surprising, however, that Wheatstone was satisfied with the statement that his researches had 'already prepared the public to receive it as a project capable of practical application'. Wheatstone's established reputation gave their telegraph an air of scientific respectability which it could not have had from Cooke alone, but Wheatstone had succeeded in making an instrument which worked through long lengths of wire when Cooke's would not. If Wheatstone had expected the award to be published, he would have required a much fuller statement of his own contributions. He was satisfied with the arbitrators' statement because it brought what he really wanted – the opportunity to continue his researches in peace.

The arbitration proceedings apparently brought to an end, for a time at least, the quarrel between the partners, and they continued to develop their telegraphs. The original partnership deed provided that further inventions by either partner should be the property of the partnership, but this agreement appears not to have been enforced by either of them. In his case to the arbitrators Cooke stated that notwithstanding their original agreement he had agreed in November 1839 that Wheatstone should be free to employ staff to manufacture telegraph instruments of his own design, for his own profit and bearing Wheatstone's name alone. In 1838, as we have seen, Cooke had obtained a patent in his sole name for a variety of minor inventions relating to telegraphs and there is nothing to suggest that Wheatstone objected. Cooke also obtained another, similar patent in 1842.

Printing telegraphs

Wheatstone's own patent of 1841 was largely concerned with electric motors and generators, and has already been mentioned, but it also included three items of telegraph interest. None of the three were adopted in practice at the time, but all are significant for what they reveal of Wheatstone's ideas and experiments at that period. The first is the printing mechanism of Wheatstone's earliest printing telegraph (see Figure 11.3a and 11.3b).

The printing telegraph is essentially a development of the ABC telegraph receiver, powered by a clockwork motor and controlled through an electromagnetically operated escapement. The rotating letter dial, or pointer, of the

Figure 11.3a   Wheatstone's printing telegraph of 1841.
This view shows the instrument assembled and ready to receive a message which will be printed on the paper wrapped around the horizontal cylinder.

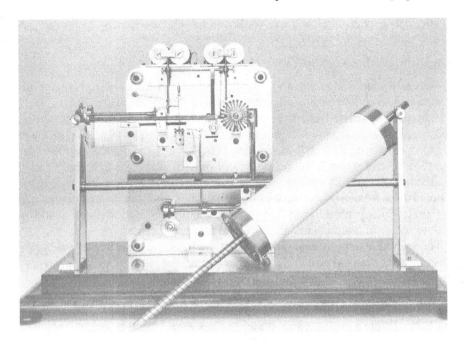

Figure 11.3b   *Wheatstone's printing telegraph of 1841.*
In this view the cylinder is detached to show the typewheel which has
24 radial arms each carrying a printer's type. The hammer which strikes
the type on to the paper can be seen at the right of the typewheel. The
instrument contains two distinct mechanisms. One is a simple 'ABC'
telegraph receiver which turns the typewheel so that the desired character is
in front of the hammer. The second mechanism pulls the hammer back
against a spring, then releases it so as to strike the type; it then turns the
drum through an angle corresponding to one letter space. As the drum turns
it advances along a screwed rod, so arranged that while a line is being printed
around the drum it advances by a line space.

ABC instrument is replaced by a 'typewheel', a small wheel with 24 radial arms. The arms are springy and each carries a printing type just clear of the paper on which the message is to be printed. The paper is wrapped round a cylindrical drum and covered with a sheet of carbon paper. In operation the typewheel is rotated until the desired type is at the right position over the paper. Then a second mechanism, also driven by clockwork, is started by withdrawing a detent electromagnetically. A weighted hammer is pulled back and then released, striking the type at the printing position so that the desired character is printed through the carbon paper. The drum then turns through an angle corresponding to one letter space, so that the message is written around the circumference of the drum. The drum is carried on a screwed rod so that it moves axially by a line space in the course of each revolution, giving sloping lines of print.

There is a printing telegraph of this kind in the Science Museum. The 24 characters on its typewheel are

ABCDEFGHIKLMNOPQRSTVWYZ*

It was an ingenious machine and it really worked, but it was never brought into commercial use. The printing telegraph was an invention ahead of its time; it was made before the basic idea of communication by electric telegraph had become generally accepted. It would have been very expensive compared with the simpler telegraphs, and it would have been slower since the hammer mechanism took several seconds to operate and print each letter whereas a man receiving the message would note the signals as quickly as they could be transmitted. Also, if there were no operator watching the incoming message, any error which put the receiving mechanism out of synchronism with the sender would go undetected until the recipient came to read his message later.

The second invention relevant to telegraphy in Wheatstone's 1841 patent was an electro-mechanical switching arrangement for providing that signals from a single circuit go sequentially to two or more local circuits. Possibly Wheatstone had the idea of using such an arrangement with his printing telegraph so that the telegraph wire would be connected first to the 'ABC' mechanism and then to the hammer mechanism. In this way the printing telegraph could have been operated through a single circuit. The instrument now in the Science Museum bears no evidence of ever having been operated other than through two distinct circuits.

The final invention in the patent was the idea of recording the time at which some action took place. A sheet of paper with a time scale on it was moved uniformly by clockwork. The action whose timing was to be recorded was made to close a contact in a circuit so as to energize an electromagnet which attracted an armature to press a pencil on to the paper. This appears to have been the first proposal for an electric recorder of the time of an event. It only recorded 'on' or 'off', so it was not yet a recording measuring instrument. Wheatstone did make recording measuring instruments at this time or shortly afterwards in connection with his meteorological interests. These instruments are considered below.

Cooke and Wheatstone had one further joint patent before the partnership finally came to an end. This was granted on 6 May 1845. The specification was mainly concerned with detailed improvements in the ABC Telegraph. Other features included the idea of a twin-needle telegraph with two needles influenced by a single coil but with their movement limited by stops so that one needle could move only to the left and the other needle only to the right. No reason was given why such an arrangement should be preferable to a single-needle telegraph whose needle could move in either direction. A possible advantage would be faster operation since, when the current ceased in the twin-needle telegraph, the deflected needle would be returning to a rigid stop whereas the needle of the single-needle instrument would tend to oscillate about the central position. A further feature was that the deflection of the needle should produce a sound, the sound being different for left-hand and right-hand deflections. The advantage of this would be that the person receiving the message could hear the signals, rather than see them, and could therefore write down the message as it was received. The idea of listening to the sound of the needles rather than watching them may well have arisen in use when an operator found that the two needles of his instrument made different sounds. It is said that the Morse sounder was developed after Morse's operators found that they could 'hear' the message being received on a Morse inker.

The Electric Telegraph Company

The ABC telegraphs described in the 1840 patent were not a commercial success then, but 20 years later Wheatstone developed more advanced versions which were sold in large numbers. During the 1840s a market developed for the simple telegraphs with one or two needles. In late 1842 the Great Western Railway agreed to let Cooke extend the telegraph from West Drayton to Slough and as part of the new agreement Cooke obtained the right to transmit messages for the public. In 1843 the telegraph was extended from Slough to Windsor and this extension proved important in bringing the telegraph to public notice. *The Times* announced the birth of Queen Victoria's second son at Windsor on 6 August 1844 and acknowledged 'the extraordinary power of the Electro-Magnetic Telegraph' by which it had received the news. Further publicity came in January 1845 when the murderer John Tawell was seen to board a train at Slough. A description was telegraphed to Paddington where the police met the train. In 1844 the Admiralty, which had so flatly rejected Ronalds' plans 30 years earlier, entered into a contract with Cooke for a telegraph between the Admiralty in London and Portsmouth. This was the longest telegraph line so far.

The technical problems connected with the telegraph were gradually solved but when it began to be exploited commercially a new problem arose. The purchase and installation of the wires in particular was very expensive. The telegraph needed far more capital than Cooke and Wheatstone could raise as a partnership, even if Wheatstone had wished to remain an active business partner. In April 1843 Wheatstone and Cooke negotiated a new agreement between them under which Wheatstone assigned his share in their patents to Cooke in exchange for a royalty payment based on the length of telegraph line completed each year. The agreement included a statement that it was made in order to simplify the business arrangements: up until that date all contracts had to be approved by both partners. All outstanding claims of one partner against the other were cancelled and the royalty was fixed at £20 per mile of telegraph completed during the year for the first 10 miles, reducing in stages to £15 per mile after 50 miles.

Cooke's records give some idea of the sums involved in setting up telegraph lines. Capital expenditure between 1836 and 1844 was £31,000 and in 1845 alone £71,000. During that period the income was £97,000. In the nature of the telegraph business it was necessary to build a extensive system before it could be economically viable. The only way to raise the necessary capital for further telegraph lines was to form a company. With this object in view Cooke sought financial backing to buy Wheatstone out completely. In August 1845, when the formation of a company was being considered but before Cooke had a financial backer, Wheatstone stated that he would commute his royalty on all lines in England, Wales, Scotland, Ireland and Belgium for a total sum of £30,000. This was agreed after some discussion together with another agreement which permitted Wheatstone to use the patented apparatus freely on lines not exceeding half a mile in length.

Cooke obtained the support of John Lewis Ricardo (1812–1862), a financier involved with several railway companies and Member of Parliament for Stoke on Trent, whom he first met on 1 October 1845. It appears that Wheatstone and Ricardo did not meet until 15 January 1846. Ricardo was very impressed with the telegraph and soon came to an agreement with Cooke. He was an able administrator and took a leading part in the promotion of the Electric Telegraph Company, to which the patents were assigned in 1846. Ricardo was chairman for ten years and during that period he introduced the use of franked message papers and the employment of female clerks. The rising telegraph industry provided new and respectable employment opportunities for women across the country.

In his *Answer* Wheatstone referred to the transaction by which he had assigned his rights to Cooke and Cooke's subsequent sale of the patent rights to the company:

The negotiations had arrived at this stage by means of a correspondence between Messrs. Cooke and Wheatstone, from which it appeared that Mr. Cooke was about to transfer both their interests in the patent rights, though he did not think it necessary to communicate to Mr. Wheatstone that the price which he was about to receive for them was £150,000.

Cooke's response was to state that, after allowing for various expenses, he had only really received about £90,000. Details of the transactions and some of the partners' letters at the time are set out in their pamphlets. Cooke raised the matter in his first pamphlet to show how 'very liberal' he had been to Wheatstone. Wheatstone clearly thought Cooke had not been as candid and liberal as might reasonably have been expected, but he did not seek to renegotiate the terms.

There was a suggestion that Wheatstone should be engaged as the Company's scientific adviser. Cooke's version of the affair was:

... the promoters of the Electric Telegraph Company especially requested me to engage him as the Company's scientific adviser and assistant, on very liberal terms; and a memorandum to that effect was signed, and for a time acted on.

Mr. Wheatstone, however, soon resigned his appointment, under the following circumstances:

A Bill for the incorporation of the Company, which was brought into Parliament in the session of 1846, was opposed by Mr. Alexander Bain, who asserted in his petition that he had invented an electric clock, and an electric printing telegraph, and had communicated his inventions confidentially to Mr. Wheatstone, and that the latter had claimed them as his own. The Directors carried their Bill, notwithstanding this opposition, though not without difficulty, through the House of Commons; but Mr. Bain's statement and evidence made such an impression in the House of Lords, that, in the afternoon of the third day of the sitting of the Lords' Committee, the Duke of Beaufort, as Chairman, intimated to the Company's counsel that the Committee were of opinion that the Company ought to make an arrangement with Mr. Bain – hinting, in fact, pretty plainly, that their Bill might be thrown out if they declined to do so. After a consultation with counsel, it was considered necessary to give way. Mr. Bain was accordingly bought off, and became associated with the Company, to the extreme displeasure of Mr. Wheatstone.

About the same time, the Directors unluckily made an agreement with a Mr. Henry Mapple, in ignorance that this person had a similar controversy with Mr. Wheatstone respecting an improved alarm and a telegraphic rope.

Wheatstone's version was rather different. He stated that the suggestion that he should be engaged as scientific adviser was not in fact adopted. It appears to have been the intention that Wheatstone would be the chairman of a committee of three, the other two being unnamed. One of the duties was to superintend submarine telegraph experiments at Portsmouth. Wheatstone stated that he acted upon the arrangement for a time without receiving a fee, though he was repaid for some expenses in labour and materials. The Mr. Mapple had been employed by Wheatstone first in the construction of telegraph instruments and then in the manufacture of submarine cable. When Wheatstone had to be away on one occasion the company wanted some cable prepared quickly and Wheatstone arranged for Mapple to supervise the work. On his return Wheatstone found that the company had arranged with Mapple for him to take out a patent for certain improvements made while carrying out Wheatstone's instructions. 'From these circumstances, combined with others', wrote Wheatstone, 'it was obvious that there was no intention on the part of the Company to fulfil Mr. Cooke's engagement of Wheatstone as scientific adviser'.

Henry Mapple's patent was obtained on 27 October 1846 and the specification included 'a method of coating the conducting wires with a metal covering by enclosing them (already insulated with cotton, &c.) in a leaden tube, reducing the tube in size, and drawing it down with grooved rollers ... '. This cable construction was not a success.

Nevertheless, Wheatstone's connection with the company continued for some time, for he wrote later:

My connexion remained, for some time after the arrangements referred to, on the same footing as before, that is, I continued always ready to give any assistance required of me, and I was on several subsequent occasions asked to do so. My connexion in this way did not cease until the beginning of 1850, though my appointment as Scientific Adviser was never confirmed.

Cooke stated, in the long quotation above, that the main reason for Wheatstone's not being engaged by the company was the fact that the company had been obliged by the House of Lords' committee to come to terms with Bain in order to get their Bill passed. The official Minutes of the Lords' proceedings simply record that evidence and argument were heard on 5 and 6 June 1846 and the hearing then adjourned until Monday 8 June when counsel for the Bill announced that Bain had agreed to withdraw his opposition and assign his own patents to the company for £7500. Bain was also to be employed by the Company to manage the manufacture of clocks. The Bill was passed.[1]

W.T. Jeans, writing after Wheatstone's death, gave an account of a dispute between Wheatstone and Alexander Bain in the early 1840s. On 26 November 1840 Wheatstone read a paper to the Royal Society describing his electro-magnetic clock which he said was his own invention, and the clock was demonstrated in the Library. According to Jeans, in the following January Wheatstone

received notice from a Mr. Barwise, of St. Martin's Lane, that he claimed to be the inventor of the clock, and shortly afterward it was stated in placards that Messrs Barwise and Bain were the joint inventors. At first Professor Wheatstone took little notice of the attacks thus made upon his originality, but in June, 1842, he was directly charged by Mr. Bain in the public press with appropriating his inventions.

The only newspaper of the period which has an index is *The Times* but there is nothing about the affair there. Jeans continues:

In reply to that accusation, Professor Wheatstone stated that Alexander Bain was a working mechanic who had been employed by him between the months of August and December, 1840: and to the allegation that Bain communicated the invention of the clock to him in August 1840, he answered that there was no essential difference between his telegraph clock and one of the forms of his electro-magnetic telegraph, which he had patented in January 1840: that the former was one of the numerous and obvious applications which he had made of the principle of the telegraph, and that it only required the idea of telegraphing time to present itself and any workman of ordinary skill could put it into practice – in telegraphing messages the wheel for making and breaking the circuit was turned round by the finger of the operator, while in telegraphing time it was carried round by the arbor of a clock. He also stated that, long before the date specified, he mentioned to many of his friends how the principle of his telegraph could be applied 'to enable the time of a single clock to be shown simultaneously in all the rooms of a house, or in all the houses of a town connected together by wires'. The accuracy of these statements was verified by Dr. W.A. Miller, of King's College, and by Mr. John Martin, the eminent artist. The latter stated that Professor Wheatstone explained to him in May 1840, his proposed application of his electric telegraph for the purpose of showing the time of a distant clock simultaneously in as many places as might be required. Mr. Martin, on hearing the explanation, said to him, 'You propose to lay on time through the streets of London as we now lay on water'. Mr. F.O. Ward, a former student of King's College, stated that Professor Wheatstone explained the matter to him on June 20, 1840. While watching the motions of the dial telegraph as he turned the wheel that made and broke the circuit, Mr. Ward remarked that if it were turned round at a uniform rate, the signals of the telegraph would indicate time, to which Professor Wheatstone replied: 'Of course they would, and I have arranged a modification of the telegraphic apparatus by which one clock may be made to show time in a great many places simultaneously'; and the Professor showed him drawings of an apparatus for that purpose, in which the making and breaking of the circuit by the alternate motion of the pendulum of a clock, would produce isochronous signals on any number of dials, provided they were connected by wire. The electric clock in question has been repeatedly tried, but has not answered expectations.

Mr. Alexander Bain also accused Professor Wheatstone of appropriating his printing telegraph. He said he communicated the invention of the electric clock, together with that of the electro-magnetic printing telegraph, to Professor Wheatstone in August, 1840, before ever Professor Wheatstone did anything in the matter. To that the Professor replied that the printing apparatus was merely an addition to the electro-magnetic telegraph, of which he was undoubtedly the inventor. As to the way in which this telegraph printed the letters, he explained that for the paper disc (or dial) of the telegraph, on the circumference of which the letters were printed, he substituted a thin disc of brass, cut from the circumference to the centre so as to form twenty-four radiating arms on the extremities of which types were fixed. This typewheel could be brought to any desired position by turning the commutator wheel. The additional parts consisted of a mechanism which, when moved by an electro-magnet caused a hammer to strike the desired type – brought opposite to it – against a cylinder, round which were rolled several sheets of thin white paper along with the alternate blackened paper used in manifold writing. By this means he obtained at once several distinct printed copies of the message transmitted. He maintained that the plan was begun and carried out solely by himself, and Mr. Edward Cowper stated, as corroborative evidence, that on June 10, 1840, he sent a note to Professor Wheatstone (who had previously told him of the contrivance by which his telegraph could be made to print), giving him information, which he had asked for, respecting the mode of preparing manifold writing paper, and the best form of type for printing on it.[2]

W.T. Henley (1814–1882), the founder of W.T. Henley's Telegraph Works Co Ltd, began his electrical career by working for Wheatstone. He left school at the age of 11 and worked in various unskilled jobs. He then taught himself science from books and by conducting experiments, and eventually established himself as a scientific-instrument maker. In that capacity he came to the notice of Gassiott, who introduced him to Daniell and Wheatstone, and he worked with Wheatstone in his telegraph experiments up to the formation of the Electric Telegraph Company in 1846. Thereafter Henley established his own business manufacturing telegraph instruments and cables.

The establishment of the Electric Telegraph Company was the final parting of the ways for Cooke and Wheatstone. Cooke was a director of the company for most of the remainder of his life. He resigned from the board in November 1849 after a disagreement (of which no further record survives) but he was re-elected shortly after. With the practical experience he had gained as the business manager of the partnership he excelled in the work of managing the company. He made no more inventions in connection with the telegraph, but 20 years later he had a new interest and obtained a series of patents for quarrying and stone-cutting machinery. He moved to Wales and acquired a quarry in Merionethshire but the new interest proved a financial disaster and he lost all the fortune the telegraph had brought him.

Neither Cooke nor Wheatstone left a record of how they felt when the partnership ended. Cooke's series of lengthy letters to his mother had ceased when he married in 1838, so we have no account of his actions at that time to compare and contrast with the early meetings when Cooke wrote about every detail and seemed to be seeking his mother's approbation for all he had done. Only one of Wheatstone's letters reports his own feelings about anything: in 1843 he told Ronalds in a letter that his work had been held up by the necessity of replying to some of Bain's attacks, 'a most grievous waste of time'. Cooke and Wheatstone were probably both relieved when they became free to go their separate ways. Wheatstone had plenty of other interests at the time, including the problems of submarine telegraphy, electrical recording of meteorological observations and electrical chronoscopy. He had also obtained another patent for musical instruments, married and moved house. With so much else to do he was probably glad to part from Cooke and the simple telegraph which was then ready for commercial exploitation. Wheatstone made a lot of money from his association with Cooke, even though Cooke made more. We do not know how much Wheatstone had invested but Cooke himself stated that 'pecuniary considerations' had been a factor when he entered into partnership with Wheatstone, and Wheatstone stated that he had had to divert into the partnership funds intended for other research. The object of this research was an underwater telegraph cable.

Notes

The main sources for this chapter are the same as for Chapter 10.

1 House of Lords Record Office: *House of Lords, Manuscript Evidence*, 1846, volume 12.
2 The quotations from W.T. Jeans are from his *Lives of the Electricians*, 1887.

Chapter 12

Submarine telegraphy

Wheatstone was interested in the possibility of a submarine telegraph from a very early stage in his work on electric telegraphy.[1] Among his papers in King's College is the following undated manuscript:

Scarcely any current passes unless water be decomposed; with platina electrodes it requires at least three of Daniell's cells to decompose water; with copper electrodes less. It is probable that if wires were laid unprotected in water if the electromotive force were not sufficiently strong to decompose water there would be no loss from want of insulation. Coating the wires with platinum or gold would if this be correct enable us to employ a stronger electromotive force without any loss from interposed moisture.

We know two cells will not decompose water even with the shortest length of wire. Have three coils each equal to two miles of wire; interpose between each two cells; will the entire circuit decompose water when the circuit consisting of one coil and two cells [will] not do so?

Most of the papers in the group from which the above quotation was taken appear to have been written about the time of Wheatstone's appointment to the College in 1834. His friend Faraday deduced the basic laws of electrochemistry during 1833 and Wheatstone would undoubtedly have taken an interest in his research; this note could well have been written soon afterwards. Faraday was interested in the relationship between the quantity of electricity which passed and the quantity of electrochemical effect.

The question implied in Wheatstone's first paragraph – whether below the threshold voltage for electrochemical action water is a good insulator – was a reasonable question to ask. Doubtless he tried the experiment and found that it was not a good enough insulator to permit submarine telegraphy with bare wires. The question in Wheatstone's second paragraph reveals that he did not, at the time of writing, appreciate that an electrolytic cell has a fixed threshold voltage below which no action takes place irrespective of the circuit resistance.

The earliest published record of his interest in submarine telegraphy occurs in the minutes of the House of Commons Select Committee on Railway Communication. On 6 February 1840 Sir John Guest, MP asked Wheatstone, 'Have you tried to pass the line through water?' The answer was, 'There would be no difficulty in doing so, but the experiment has not yet been made'. The Chairman, Lord Seymour, asked, 'Could you communicate from Dover to Calais in that way?', to which Wheatstone replied, 'I think it perfectly practicable'.[2]

In an appendix to his *Answer* to Cooke's pamphlet in 1846 Wheatstone referred to his evidence to the House of Commons Committee and said that Sir John Guest 'was previously acquainted with his plans'. It is not now clear, however, what Wheatstone's plans were at that stage. In his *Answer* he continued, writing in the third person:

Shortly after this, having been furnished with the necessary hydrographic information by his friend Sir Francis Beaufort, and received much useful counsel from the late Captain Drew of the Trinity Board, Captain Washington, and other scientific naval friends, he prepared his detailed plans, which were exhibited and explained to a great number of visitors at King's College, among whom were the most eminent scientific men and public authorities. He also made the subject known in Brussels . . .

Mr. Wheatstone's plans were also shown in 1841 to some of the most distinguished scientific men in Paris, who came to see his experiments at the College de France.

One of Cooke's letters to his mother, written in June 1837, refers to a 'rope' which had just been completed by Enderbys Brothers. This was apparently an insulated cable. In his 'Case' to the arbitrators Wheatstone claimed that before meeting Cooke he had:

made arrangements for trying an experiment across the Thames, from my lecture-room to the opposite shore. Mr. Enderby kindly undertook to prepare the insulating rope containing the wires, and to obtain permission from Mr. Walker to carry the other termination to his shot-tower. After many experiments had been made with the rope, and the permission granted, I relinquished the experiment; because after my connexion with Mr. Cooke, it was necessary to divert the funds I had destined for this purpose to other uses.

The expression 'across the Thames' does not necessarily mean that the telegraph line passed under water, but the reference to an 'insulating rope' implies that that was the plan. Cooke hotly contested Wheatstone's statement above, and produced correspondence proving that he had conducted business with Enderbys. That did not prove, of course, that Wheatstone had not had dealings with Enderbys. When the partnership was formed Cooke as manager might well have taken over the correspondence with the firm. Cooke's comments on Wheatstone's statement make it clear that the insulating rope was intended to be waterproof.

The 'detailed plans' for a submarine telegraph referred to by Wheatstone were not published in his lifetime though they were shown to his scientific acquaintances. The original drawings, which were prepared by a Polish draughtsman, A. Lutowski, are in King's College Library. In February 1876, four months after Wheatstone's death, his son-in-law Robert Sabine wrote to the Society of Telegraph Engineers (now the Institution of Electrical Engineers) informing them of Wheatstone's early plans on the subject (see Figure 12.1), which, he said, were

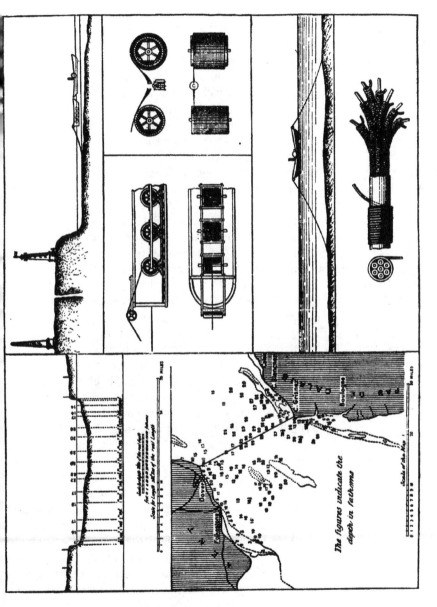

Figure 12.1 Wheatstone's plans for a submarine telegraph across the English Channel, published in the Journal of the Society of Telegraph Engineers after his death, but drawn many years earlier, probably in the late 1840s.

completed in October 1840 and shown to visitors at the College. The letter was printed, together with a copy of the drawings, in the journal of the Society. Each wire was to be the core of a rope well saturated with boiled tar. The detailed drawings leave no doubt that Wheatstone had given considerable thought to the practical problems of making and laying a submarine cable. He also attempted to foster interest in Belgium and France in the idea of a cross-channel telegraph, and he conducted a number of practical submarine trials.

In a letter written to Cooke on 20 January 1838 about various matters connected with their telegraphs, Wheatstone wrote 'I start this evening for Ostend but cannot take the apparatus with me, which is to be sent via Antwerp as soon as the ice will permit'. Presumably this was a visit to interest his continental acquaintances in the telegraph. In early 1840 Wheatstone demonstrated the five-needle telegraph in Paris. There is no evidence that submarine telegraphy was mentioned then, but it certainly was when he visited Brussels later the same year. In October his friend Professor Quételet gave an outline account to the Académie Royale des Sciences et Belles-Lettres de Bruxelles of demonstrations given at the Royal Observatory in Brussels and said that Wheatstone had found a way of transmitting signals between England and Belgium 'malgré l'obstacle de la mer'. No details of the instruments were given in order not to prejudice Wheatstone's patent rights. The demonstration and proposal for submarine telegraphy were also reported in the Brussels newspaper *Le Fanal* of 30 September 1840. In 1841 Wheatstone was again in Paris, this time showing his detailed plans. A copy of these was made by J. Joseph Silberman and given to Pouillet, who deposited them in the Conservatoire des Arts et Métiers. The copy is no longer extant, because all Pouillet's papers were destroyed in the revolution of 1848.

The agreement of 12 April 1843 by which Wheatstone assigned most of his patent rights to Cooke in exchange for a royalty did not 'extend to prevent the said Charles Wheatstone from establishing electric telegraph communication between the coasts of England and France ... for his own exclusive profit'. Wheatstone never made use of that clause in the agreement, but the fact that it was written into the terms suggests strongly that at that stage Wheatstone was interested in submarine telegraphy and saw a future in it, while Cooke was more interested in the inland telegraphy.

In June 1843 Prince Albert visited King's College. The occasion was the opening of a 'museum' of philosophical instruments and astronomical apparatus. This was the George III Collection of scientific instruments (now in the Science Museum), which the King had had made for the education of his sons. Queen Victoria presented the collection to the College, who were placed in a difficult position. They could not decline the generous royal offer but the cost of displaying the collection was an embarrassment. Wheatstone was in charge of the collection, which was displayed in a 'spacious and well-lit apartment'. After the formal opening the Prince was shown various demonstrations and then went to the terrace in front of Somerset House, where according to *The Times*:

It had been intended that some experiments should be tried by the electric fluid from the top of the shot manufactory across the river to the terrace, but the experiment was not exhibited, as it was understood that the string or wire had been cut or broken by which the communication was to be effected.[3]

There is an engraving of the event in the *Illustrated London News*, from which it can be seen that the ABC telegraphs exhibited were Wheatstone's own. From the account in *The Times* it seems that the experiment was a failure on the day Prince Albert came. Perhaps the 'string or wire had been cut or broken'; perhaps the insulation failed. However, since the demonstration had been included in the royal tour it seems reasonable to suppose that it had worked previously. The *Illustrated London News* account does not explicitly state that the wire went underwater, though if it did not do so the demonstration would not have been very remarkable at that date.

Trials in Swansea Bay

In August 1844 Wheatstone conducted some submarine telegraphy experiments in Swansea Bay. Nothing was published about this until after his death, when the experiments were brought to notice by Sabine's letter to the Society of Telegraph Engineers. Wheatstone stayed with L.W. Dillwyn, whose manuscript diary in the National Library of Wales at Aberystwyth confirms Sabine's account. Dillwyn, a Welsh MP and a Fellow of the Royal Society, already knew Wheatstone. He enjoyed scientific company and had prominent scientists visit him from time to time at his Swansea home. The entries in Dillwyn's diary about Wheatstone's visit are all very brief. Wheatstone arrived on 13 August 1844 but until the 26th Dillwyn recorded only that Wheatstone made a visit to the Royal Institution of South Wales in Swansea, went to church and took what were apparently country walks and drives. On the 26th Wheatstone went to see the Hafod Copper Works, owned by Dillwyn's neighbour. On the 27th the whole entry reads:

Mr. Wheatstone having placed a Bell in our dining room with connecting wires to a magnet in the lodge explained the principle of his telegraph to George Lennox, Dr. Newsom, Mr. Benson, and several others who I had invited to luncheon.

On the next day Wheatstone, Dillwyn and John, Dillwyn's son, 'made experiments on the passage of the electric current through sea water' from a boat to an island. On the following day Wheatstone and John continued the experiments. On 2 September Wheatstone returned to London.

Why did Wheatstone go to Swansea? He already knew Dillwyn so he may have been invited for a holiday, and Dillwyn might well have suggested that he should bring a few telegraph instruments to demonstrate. The idea of submarine cable experiments in the bay could have arisen in after-dinner conversation about the future of the telegraph. Perhaps Dillwyn's neighbour offered to make a length of cable. Wheatstone's visit to the copper works would then have been to inspect and collect the cable, which he tried out the next day. There is no indication of the length but it need only have been a few metres. Wheatstone could have gone to Swansea for secrecy, not to avoid Cooke since their agreement at that time permitted Wheatstone to work on submarine telegraphy, but perhaps to avoid publicity which might invalidate a patent, or to avoid ridicule in the event of a failure. If, however, Wheatstone had gone to Swansea with the prior intention of making some underwater experiments with a locally made cable there ought to have been additional visits to the works mentioned in Dillwyn's diary.

The 1840 plans envisaged a cable insulated with rope soaked in tar. Mapple's patent of 1846, which has been mentioned above in connection with the Electric Telegraph Company trials at Portsmouth, does not specify the insulation around the conductor, but envisages the outer lead tube being wrapped in rope and passed through a bath of hot pitch. In the early part of 1846 Wheatstone paid Mr W.H. Darker of 9 Paradise Street, Lambeth 'to enclose a copper wire insulated with worsted and marine glue in a lead tube'. The insulation used at Swansea is not known, though it was probably either rope and tar or worsted and marine glue. Neither of these was a success, however, and practical submarine telegraphy had to await the introduction of gutta percha insulation, extruded without any joint onto the conductor.

The first submarine telegraph cable which might reasonably be called a success was laid across the English Channel in 1851 by the brothers J. and J.W. Brett, and T.R. Crampton. Wheatstone was not personally involved, nor did the Electric Telegraph Company play any part. Wheatstone took only a minor part in the commercial development of submarine telegraphy, but two pieces of scientific work may be mentioned in that context.

The capacitance of cables

During May and June 1854 he conducted a series of experiments at Greenwich on 110 miles of six-core cable which had been made for laying in the Mediterranean from Italy to Corsica. The experiments are summarized in a paper in the *Proceedings of the Royal Society* for 1855. This paper did not present any new theoretical conclusions, but was a confirmation and continuation of some work of Faraday's. One fact Wheatstone did establish in the paper was that a cable connected to a battery 'becomes charged to the same degree of tension throughout its entire extent', however long the cable might be. Submarine telegraph cable had electrical characteristics unlike anything previously known. The conducting wire, surrounded by a thin layer of insulating material and then by a conducting layer of sea water, became in effect a very large capacitance. (Wheatstone used the term 'Leyden jar' in his second paper on the subject, though he avoided any such term in his first. Faraday used the term 'Leyden battery' in his paper.) The cable was not exactly analogous to a Leyden jar because the capacitance was distributed along the length of the wire and the wire had appreciable resistance. If a voltaic battery was connected between the conductor and the surrounding water, then the cable became charged like a Leyden jar, but an appreciable time would elapse before the cable was fully charged. If the battery were then removed and the cable discharged the process again took time. Faraday outlined these phenomena and described the effect of discharging a charged cable by touching it and receiving a shock.

The shock was rather that of a voltaic than of a Leyden battery: it occupied *time*, and by quick tapping touches could be divided into numerous small shocks: I obtained as many as **40** sensible shocks from one charge of the wire.

In 1854 the subject was mainly of academic importance, but with the introduction of Wheatstone's automatic telegraph in 1858 it was to become of practical importance also. The time required to charge a cable imposed a fundamental limit on the speed with which a message could be sent. It was appreciated even in 1854, however, that it was not necessary to charge and discharge the whole cable for each signal. Faraday described experiments using the several underground telegraph wires running between London and Manchester connected to give a total length of 1,500 miles of cable with both ends and several intermediate points all together in London. He stated that if a battery were connected briefly at one end of the wire, and the current were monitored at one end, at the mid point, and at the far end of the cable by three galvanometers *a, b, c*, placed close together, it was possible that instrument *a*:

could be deflected and could fall back into its neutral condition, before the electric power had reached *b*; which in its turn would be for an instant affected, and then left neutral before the power had reached *c*; a wave of force having been sent into the wire which gradually travelled along it, and made itself evident at successive intervals of time, in different parts of the wire. It was even possible, by adjusted touches of the battery, to have two simultaneous waves in the wire following each other.

The submarine cable inquiry

Wheatstone was a member of the Committee of Inquiry appointed after the failure of the attempts in 1857 and 1858 to lay a telegraph cable across the Atlantic. The second cable, laid in 1858, lasted a few weeks and showed that transatlantic submarine telegraphy was a real possibility. The British government had agreed to guarantee some of the cost of the unsuccessful cables and had provided a ship, *HMS Agamemnon*, for the laying operation. The government also had a financial interest in a cable laid in the Red Sea from Suez to Aden, and thence to India. This was also a failure, and before spending any more public money on such projects it was decided that a committee should be appointed to investigate the whole question of the construction, laying and maintenance of submarine cables. The Committee had eight members. Four were appointed by the Privy Council Committee for Trade: George Bidder, William Fairbairn, Douglas Galton and Charles Wheatstone. The Atlantic Telegraph Company appointed Edwin Clark, Latimer Clark, George Saward and Cromwell F. Varley. Robert Stephenson was initially a member of the

Committee, but he died just after it began work. They sat from December 1859 to September 1860, questioning many people with experience of submarine work and making their own investigations, and reported in April 1861.[4]

Wheatstone's chief contribution to the Committee of Inquiry was a paper 'On the Circumstances which influence the Inductive Discharges of Submarine Telegraphic Cables'. It was a long paper, about 14,000 words, and was clearly the outcome of extensive research. He tried a variety of cable samples. Some had copper cores and some iron; some had gutta percha insulation but other materials were used also; some of the samples differed in the diameter of the core or the thickness of the insulation. The cables were all coiled and placed in a tank filled with water. He charged the test cables and monitored the discharge when they were earthed through a galvanometer, and he studied the effect of length, core diameter, insulation thickness, conductor resistivity, pressure and temperature on the discharge.

With short lengths of cable the capacity is small and the discharge current difficult to measure. In order to measure the capacitance of such cables Wheatstone constructed an instrument which he called the 'accumulating discharger'. This was a two-way switch which alternately connected the cable to a battery and to the galvanometer, so that the cable was successively charged and then discharged through the galvanometer. The switch was operated through a cam turned by hand and could perform 68 switching cycles per second. Since the galvanometer had a time constant of several seconds it gave a reading corresponding to several hundred 'accumulated' discharges.

At the end of his paper for the Committee of Inquiry Wheatstone referred to Latimer Clark's 'inductometer' which was similar to his own accumulating discharger. The inductometer was published in a letter to the *Engineer* of 28 September 1860. Wheatstone was apparently anxious to establish his priority, and stated that the accumulating discharger was made in April 1860 but the inductometer was not made until September of that year. Having claimed his priority over Latimer Clark, however, Wheatstone then referred to papers by C-M. Guilleman who had made a similar apparatus but for another purpose in 1849. Guilleman had connected a capacitor to a battery through a reversing switch and galvanometer, and he showed that by changing the switch a current could be taken from the battery even though there was no connection between its poles. In a later paper of October 1860 Guilleman described the application of his apparatus to the measurements of charges on telegraph cables.

This work on submarine cables may be compared with Wheatstone's earliest work on the electric telegraph. In both, his theoretical understanding of electric circuits and his proficiency in making electric measurements were applied to the solution of practical problems. This was the kind of work which established Wheatstone's scientific reputation in telegraph matters.

Wheatstone was the obvious person for the Prince Consort to consult when he had an idea for a new telegraph cable. Prince Albert kept in touch with the progress of science, including the successes and failures of various submarine telegraph trials. In December 1860 the Royal Librarian at Windsor wrote on behalf of the Prince to Wheatstone to seek his view on an idea which had occurred to the Prince and which might ameliorate the problems of submarine telegraphy:

Windsor Castle, Dec^{br} 17th 1860

Sir,

I am commanded by His Royal Highness the Prince Consort, to draw your attention to the following matter.

The Prince in common with everyone to whom the progress of our telegraphic communications as one of the great civilizing agents of the day, is of great importance, has often reflected upon the unsatisfactory results obtained by our submarine cables which have from their construction hitherto baffled all exertions to establish long deep-sea communications. The bulk and weight of the cables, the want of elasticity, the difficulty of transporting them and spinning them out, and perhaps above all the difficulty of their manufacture on account of the incongruity of the many different substances of which they are composed, together with the great pressure exercised by the water upon them, – may be looked upon as some of the most essential causes of failure.

In reflecting upon how some of these might be overcome the idea suggested itself to His Royal Highness, whether water itself could not be used as the conducting medium? A simple tube of cauotchouc, such as our flexible gas-tubes, might enclose and isolate a column of water, through which the spark might be sent. It would be cheap of manufacture, homogeneous in material, flexible, light, and the outward pressure would be neutralized as the fluid inside would be the same as that outside.

There may occur to you a thousand reasons, why this idea cannot be practically carried out; but His Royal Highness thought you would forgive the trouble if he asked you, who stand in such a parental position to our telegraph system, to let His Royal Highness have your opinion upon it.

I am, Sir,
Your obedient servant,
    C. RULAND
    Librarian

Wheatstone's reply does not survive, but among his papers at the College is a note that 'It would require a tube 9 feet in internal diameter for the salt water it contains to be equal in conductivity to a copper wire $\frac{1}{16}$ of an inch in diameter'. Presumably Wheatstone incorporated this information into a tactful letter to the Prince.

Notes

1 A good general account of the early submarine telegraphs is given in Charles Bright, *Submarine Telegraphs*, 1898.
2 Minutes of the House of Commons Select Committee on Railway Communication, *House of Commons Sessional Papers*, 1840 (xiii).
3 *The Times*, 23 June 1843, p. 7.
4 *Report of the Committee of Inquiry on the construction of submarine telegraph cables*, HMSO, 1861.

Chapter 13

The developing telegraph

The instruments that Cooke installed between Paddington and West Drayton were those designed by Wheatstone and they indicated letters of the alphabet directly. All the other telegraphs set up by Cooke used single- or double-needle instruments which gave coded signals. These instruments were simpler to construct and maintain but required specially trained operators. Wheatstone was always interested in direct reading telegraphs and in 1858 and 1860 he was to obtain patents for much improved ABC telegraphs which were successfully exploited by his Universal Private Telegraph Company. At the same time he developed automatic telegraph apparatus using punched paper tape to convey messages several times faster than was possible with manual operation.

In his final agreements with Cooke, Wheatstone retained two distinct concessions. One was the right to establish a cross-channel telegraph, which he did not succeed in doing. The other was the right to use the patented instruments between places not more than half a mile apart. The patents referred to in that agreement had all expired by the time Wheatstone established the Universal Private Telegraph Company so the half-mile limit never applied, but the fact that he went to the trouble of keeping the rights shows that he realized there might be a market for short-distance private telegraphs. Cooke was chiefly interested in long-distance communication.

Wheatstone's work with the Select Committee on Ordnance, which has already been mentioned, included advising on the military applications of the telegraph. Although there is no evidence that it led directly to the improved ABC telegraphs patented in 1858 and 1860 (see Figure 13.1), it may well have revived Wheatstone's interest in this type of instrument, first devised 20 years before.

The instruments described in the 1858 patent specification operate in the same way as those of 1840, but there are many detailed refinements. Several modifications of the 'escapement telegraph' are described which give a more sensitive receiving instrument, but the most important innovation in relation to the receiver is the use of a magnetized armature which is moved one way or the other according to the direction of the line current, instead of a soft-iron

Figure 13.1   Wheatstone's ABC telegraph in its most developed form with both transmitter and receiver in one unit.

armature which is always attracted whatever the direction of the current. The basic drive mechanism has two parallel, magnetized needles fixed on opposite sides of an axis of rotation, lying alongside a straight electromagnet. This 'polarized' drive mechanism gives increased sensitivity and was used in almost all Wheatstone's later telegraphs. It requires a signal in the form of alternate positive and negative pulses rather than the unidirectional pulses used in his earlier instruments. Where the transmitting instrument was battery powered, its contact mechanism was necessarily more elaborate. As in 1840 Wheatstone described two types of transmitting instrument: one had a switching arrangement and a battery, the other had a magneto. The magneto was much the same as he had used in 1840, with a pair of coils rotating alongside a horseshoe magnet and a commutator so that the pulses were unidirectional. With the 1840 telegraph the person sending a message had to start and stop the magneto for each symbol transmitted; with the 1858 arrangement the magneto was turned continuously and a contact mechanism allowed current to pass from the magneto to the telegraph line at the desired times. Since the magneto was being turned continuously it could also be turned faster than in the 1840 instruments and could therefore be smaller.

The contact mechanism of the 1858 transmitting instrument with a magneto was the invention that converted the ABC telegraph from a curiosity to a practical telegraph usable by anyone who could read. The 'communicator', as

Wheatstone called it, has a dial and pointer similar to that on the receiver. The dial is surrounded by buttons, one opposite each symbol on the dial. The buttons are linked by a chain so that only one may be depressed at a time, and when a second button is pressed the first is released. The pointer is coupled through a friction drive to the magneto, and its shaft carries a projecting catch. Each of the buttons has a lever capable of engaging the catch when that button is depressed. If the magneto is turned continuously the pointer will be turned by the friction drive until the catch meets the lever. When another button is pressed the first lever is withdrawn and the pointer-shaft moves on. Contacts are arranged so that the magneto output is short-circuited whenever the catch and a lever are engaged. In this way signals are transmitted only while the pointer is in motion towards the next symbol. The 1860 patent specification gives details of further improvements in the telegraph but the most important feature is the use of a reluctance generator in the transmitting instrument. It has the advantage of being very robust and requiring little or no maintenance.

Codes and ciphers

Wheatstone's work on electric telegraphs led him into other fields of interest. He devised ciphers so that telegraph messages could be kept secret; he made typewriters whose mechanism was developed from the printing telegraph; he designed electric clocks and chronoscopes; and he developed systems for transmitting the readings of meteorological instruments to a distant station and instruments which recorded their own readings.

Wheatstone was always especially interested in telegraph systems in which the receiver indicated letters of the alphabet directly rather than systems using coded signals such as Morse code. Sometimes it is desirable to have a telegraph which indicates letters directly so that an untrained operator can use it, and yet have the message in cipher so that only those having the key to the cipher can understand it. (The words 'code' and 'cipher' are used here in their narrow senses. In a 'code' letters of the alphabet are replaced by symbols, such as the dots and dashes of Morse code; in a 'cipher' letters are replaced by other letters.) Wheatstone devised the cipher often known 'Playfair Code'. Lyon Playfair gave an account of it at a dinner at Lord Granville's house in Carlton House Terrace in January 1854 and recorded the occasion in his Memoirs:

At this dinner . . . I explained to Lord Palmerston Wheatstone's newly-discovered symmetrical cipher, which I thought might be of use in the Crimean War then pending. It consists in taking a key word such as 'Palmerston', and writing the remainder of the alphabet symmetrically under it. The cipher to be sent consists of the letters at the opposite angles of the nearest rectangle. I told him that I had gone with Wheatstone to the Foreign Office and explained it, but the Under Secretary objected to it as being too complicated. We proposed that he should send for four boys from the nearest elementary school in order to prove that three of them could be taught to use the cipher in a quarter of an hour. The reply made to this proposal by the Under Secretary was complimentary to our diplomatic service. 'That is very possible, but you could never teach it to attachés!' I constructed an alphabet hastily with the key word 'Palmerston', and showed to Lord Palmerston and his colleagues how it could be used. The next day I went to Dublin, and while there I received two short letters in cipher, one from Lord Palmerston, the other from Lord Granville, showing that they had readily mastered the cipher.

Lyon Playfair and Wheatstone were neighbours until Playfair became Professor at Edinburgh in 1858. Playfair lived near the Surrey end of Hammersmith Bridge and Wheatstone at the Middlesex end. Playfair also recorded that:

On Sundays we generally walked together, and used to amuse ourselves by deciphering the cipher advertisements in 'The Times'. An Oxford student who was in a reading party at Perth was so sure of his cipher that he kept up a correspondence with a young lady in London. This we had no difficulty in reading. At last he proposed an elopement. Wheatstone inserted as an advertisement in 'The Times' a remonstrance to the lady in the same cipher, and the last letter was, 'Dear Charlie, write no more, our cipher is discovered!' One cipher appeared each month in 'The Times', but it was so short that it was difficult to read. At last we made it out to be, '"The Times" is the Jeffreys of the Press'. Anyone acquainted with cipher will see that the key to this short advertisement was the frequent repetition of the letter 'e' and of the Word 'the'. On telling Delane, the editor of 'The Times', that his paper was publishing its own condemnation as the wicked judge, he was angry instead of being amused at the trick played.

The 'newly-discovered symmetrical cipher' by Wheatstone which is described in Playlair's account above forms an important part of the plot of Dorothy Sayers' detective novel *Have His Carcase*. Lord Peter Wimsey interprets a cipher message, which he describes as being 'Playfair', and so solves the mystery. Lord Peter claimed to have used it in the 1914–1918 War. The 'Playfair' cipher was in fact used by the British in both the Boer War and the 1914–1918 War. Wheatstone has never received public credit for that cipher, but there is a rough justice in that, for 'Wheatstone's Cryptograph' described below was not in fact an original idea of Wheatstone's. It had been known previously as the St Cyr Wheel.

Wheatstone's Cryptograph (see Figure 13.2) was an instrument for putting a message into cipher (a different cipher from the previous one). The text of an instruction leaflet is included in the collected volume of his Scientific Papers, and several of the instruments survive. Neither the instrument nor the text are dated, but the word 'Sebastopol' is used as a keyword in an example in the text, which suggests that the text was written after the beginning of the Siege of Sebastopol on 17 October 1854.

In the introduction to his instruction leaflet Wheatstone explained the need for a cipher which was simple to use and yet truly secret:

A cipher which at the same time should be perfectly secure and easy in its application is a desideratum; and these combined advantages can only be obtained by means of an instrument in which all the complexity necessary to ensure security shall be effected by mechanical arrangements, whilst its manipulation shall be subjected to the simplest rules. Such an instrument is now offered to the public.

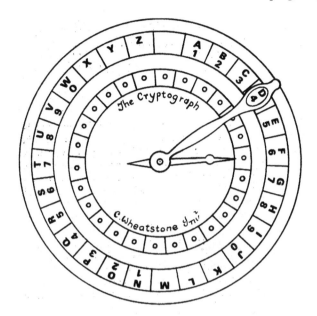

Figure 13.2   Wheatstone's cryptograph for encoding messages. The outer circle of letters is fixed. Letters may be placed in any desired order in the inner circle. The handle and pointer are geared together in the ratio 26:27.

One thing is yet wanting to render the benefits of the Electric Telegraph complete. Letters by post are sent sealed, and their contents are conveyed, with a secrecy seldom violated in free countries, to their destination; but telegraphic messages are in general transmitted so that their contents are understood by all the officials concerned in their conveyance . . .

The number of telegraphic messages relating to domestic occurrences are very much limited by the disinclination of parties to let their family affairs be known to officials in their neighbourhood; and there can be no doubt were this difficulty removed, this class of messages would be considerably augmented, to the benefit of the telegraphic department as well as to that of the public.

The advantages of communication by cipher for military purposes are too obvious to be insisted upon.

The cryptograph itself is a metal pressing ten centimetres in diameter with two hands and two circles on which letters can be arranged in any order. There are two hands on concentric axes and connected through gears giving a ratio of 26:27. The outer circle of letters has the alphabet and one blank space. The inner circle of letters has the 26 letters of the alphabet, but no space, arranged with the aid of a keyword in a manner agreed between the correspondents. The letters of the inner circle could be written on a ring of card, but for additional secrecy the 26 letters could be on 26 segments of paper which were removed from the instrument when not in use. The order of letters could only be recovered by knowing the keyword. Wheatstone suggested that if the keyword were, say,

SEBASTOPOL the repeated letters should be struck out and the remainder of the alphabet written beneath thus

```
S   E   B   A   S   T   O   P   Ø   L
C   D   F   G   H   I   J   K   M   N
Q   R   U   V   W   X   Y   Z
```

The letters for the inner circle could then be arranged by reading the columns vertically to give

SCQEDRBFUAGVHWTIXOJYPKZMLN

To use the cryptograph the long hand is first brought to the blank and the short hand in line with it. The long hand is then brought to each letter and space of the message successively around the outer circle, always moving in the same direction, and the corresponding cipher letter is indicated on the inner circle by the short hand. To decipher a message the long hand is turned so that the short hand points to the letters of the cipher on the inner circle and the original letters and spaces are read from the outer circle. The cryptograph provides a cipher which is more secure than any simple letter substitution cipher, because each letter of the message is represented by a different letter of the alphabet each time it occurs.

Wheatstone appears to have gained a reputation as a cipher expert, for in 1858 he was invited to decipher a letter for the British Museum. The letter consisted of seven pages of numbers, and every page was signed by Charles I and Lord Digbye. Wheatstone published a brief article about it, in which he explained that the cipher used two-digit numbers to represent letters and three-digit numbers to represent words in a vocabulary. He added that 'The task of translation was rendered more tedious from the original being in a different language to that in which I had first assumed it to be'. In fact the letter was in French. Wheatstone succeeded in deciphering the letter, which was concerned with the terms for a royal marriage contract.

Chronoscopes

Wheatstone's repeater clock system in which any number of 'slave' clocks were driven by one 'master' clock which has already been mentioned (see Figures 13.3a and 13.3b). He also made a chronoscope, which is a device which measures an interval of time, rather than showing the time of day. He considered both to be developments of his telegraphs.

In 1845 Wheatstone gave an account of his own work on chronoscopy up to that time in a paper published in the *Comptes Rendus* of the Paris Academy of Sciences. Wheatstone was prompted by an earlier paper by Breguet in which Breguet credited Wheatstone with one contribution to chronoscopy, but only one, and credited Konstantinoff and Breguet with an instrument which had in fact been designed by Wheatstone.

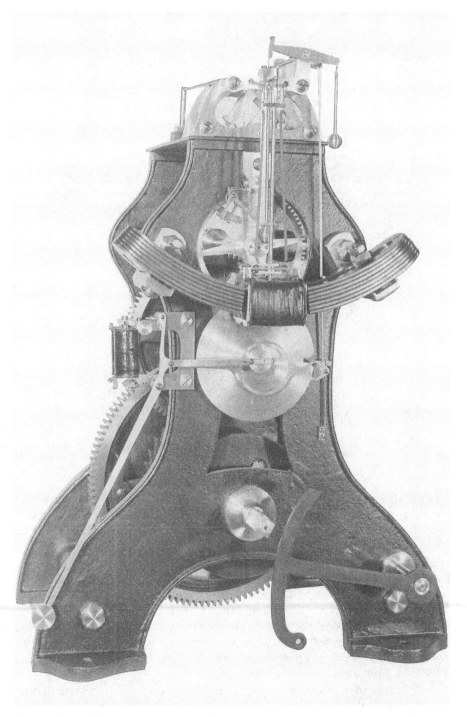

Figure 13.3a   Wheatstone's master clock a *which sent pulses to drive a number of slave clocks* b *(see page 174).*

Figure 13.3b   Wheatstone's slave clock, driven by pulses from the master clock a
(see page 173).

The earlier contribution to chronoscopy was the idea of observing, by the light of a spark, rotating paper discs with radial lines drawn on them. Wheatstone described an instrument in which three paper discs one inch across were driven by a mechanism at, for example, 2, 20 and 200 revolutions per second. A radial line was drawn on each disc and when illuminated by an electric spark they all appeared as at rest. But when they were lit by a flash lasting 1/200 of a second the third and fastest disc appeared uniformly shaded while the second showed a shaded sector of 36° (one tenth of the disc). With a flash lasting only 1/2000 of a second the fastest disc showed a shaded sector of 36°. This instrument was intended to measure the duration of flashes such as those produced by the ignition of gunpowder and can be used to measure the duration of the flash from a photographic flash gun.

The purpose of Wheatstone's paper is to claim priority and to state that he had invented at the beginning of 1840 an instrument which Breguet had attributed to M. de Konstantinoff, a Captain of Artillery in the Russian Imperial

Guard. The object of the chronoscope was, Wheatstone wrote, to measure rapid movements and above all the speed of projectiles. Wheatstone had corresponded on the subject with his friend Captain Chapman of the Royal Artillery and had had 'an interview about it with Lord Vivian, the Master General of Ordnance'. On 17 July 1841 he had explained his instrument and its uses at the Royal Artillery Institution at Woolwich. According to Wheatstone a written account of that meeting, at which 22 officers were present, was in existence when he wrote his paper in 1845, but that account cannot now be found. He also displayed the earlier rotating disc chronoscope referred to above. During the autumn of 1840 he visited Brussels and described the instrument to Quételet and the Brussels Academy of Sciences, and in May 1841 he described it and showed drawings of it to members of the Paris Academy of Sciences.

The instrument which Wheatstone stated that he had made in 1840 consisted of:

a clockwork movement driving a pointer which went or stopped in response to an electromagnet, which attracted a piece of soft iron when a current flowed in the coil of the magnet and released it when the current ceased.

Wheatstone's description (translated from his French) is not adequate to decide exactly how the instrument worked. It is not clear whether the clockwork movement included an escapement or whether it was free running. It seems very unlikely that Wheatstone would have omitted to include an escapement but later in the paper he describes another instrument which he specifically stated did have an escapement. Also it is not clear whether the whole mechanism ran or stopped under the control of the electromagnet, or whether it was just the indicating needle which started and stopped and the electromagnet controlled a clutch rather than a brake.

Whatever the precise mechanical arrangement, Wheatstone found a basic defect in the instrument: when the current ceased to flow in the electromagnet the armature was not immediately released but remained in contact for an appreciable fraction of a second. Furthermore, the period of delay was not constant, but depended on the magnitude of the current in the circuit. If the current were kept to a low value to minimize the effect, however, then the electromagnet would not attract the armature reliably and quickly at the commencement of the timed interval. He overcame this difficulty to some extent by arranging that before the start of the timed interval the electromagnet was holding the armature with the smallest possible current, and that the current was stopped at the commencement of the interval. At the end of the timed interval the electromagnet was energized by a much higher current, so as to pull the armature in rapidly. For measuring the speed of a projectile fired from a gun he arranged for the projectile first to break a fine wire across the gun barrel, which started the chronoscope by disconnecting the electromagnet, and then to strike a target which, when moved slightly, completed a circuit to stop the chronoscope. He considered that this arrangement was accurate to within one sixtieth of a second.

In the Science Museum there is a chronoscope which came from King's

College with other Wheatstone material, though there is no actual proof that it was made or designed by Wheatstone (see Figure 13.4). It is driven by a weight and has an escapement consisting of a vibrating reed engaging a toothed wheel. A pipe ending alongside the reed permits the operator to blow the reed, perhaps just to start it vibrating. The indicating needles are connected to the drive mechanism through an electromagnetically operated clutch. This particular chronoscope is probably a later development of the instruments described in Wheatstone's paper of 1845, but its exact place in the history of the chronoscope is an open question: the tuned reed escapement and the use of a clutch in a chronoscope are usually attributed to Hipp in 1848.

The fact that Wheatstone listed in such detail the people who had known about his instrument and when suggests that in 1845 he was very anxious to establish his position as inventor, although he had not patented the device. The

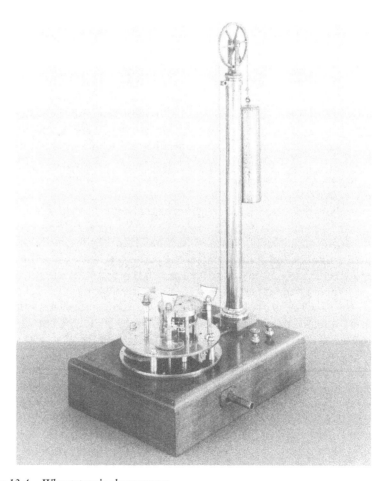

Figure 13.4   Wheatstone's chronoscope.

experiments at Woolwich in 1841 were not continued because, he said, he did not have the time and 'no-one in England ... showed any desire to continue them'. Then in 1842 he met Konstantinoff who took a considerable interest in the affair and asked for a chronoscope so that he could return to Russia and conduct experiments similar to those Wheatstone had wanted to carry out. Konstantinoff agreed not to publish a description of the chronoscope and in January 1843 Wheatstone provided him with an instrument which differed in some respects, but worked in a similar way. That instrument was presumably the one which Breguet attributed to Konstantinoff.

Wheatstone may have been particularly anxious to establish his position with regard to the development of chronoscopes because his position with regard to electric repeater clocks had been challenged by Alexander Bain. Bain worked for Wheatstone between August and December 1840, and claimed that he 'communicated the invention of the clock to him in August 1840'. In October 1840 Wheatstone wrote a Royal Society paper 'On the Electro-Magnetic Clock' of which a report, though not the full paper, appeared in the *Proceedings of the Royal Society*. The report is a simple description of an arrangement using what is virtually an early ABC telegraph receiver fitted with clock hands. This was then connected to a battery through a switching device on the scape wheel axis of a master clock. Alternatively a magneto-electric arrangement, not described, could be used. Unusually, for one of Wheatstone's publications, the report contains no references to sources, but is purely a description of apparatus.

Wheatstone himself seems to have thought that the repeater clock was an obvious application of the ABC telegraph, which any workman could put into practice once the idea of telegraphing time had been suggested, and that the idea had been suggested at least as early as May 1840.

In 1841 Bain obtained patents for electric repeater clocks which were later taken over by the Electric Telegraph Company. The clock systems were not successful, however, because the problem of keeping the contacts working satisfactorily with the small contact pressure which could be permitted proved insuperable. Wheatstone later made a repeater clock system which repeated satisfactorily but was inherently a poor time-keeper. The master clock was a pendulum clock whose pendulum carried a coil which passed over a permanent magnet. Currents induced in the coil with each swing of the pendulum drove the repeaters which were based on the later ABC telegraphs. Wheatstone presumably did not appreciate that the pendulum would experience electromagnetic forces, and that therefore the master clock would be a poor time-keeper. The system was tried by London University and the Royal Institution in 1873, but abandoned after Wheatstone's death.

Wheatstone was involved in the Observatory which the British Association maintained at Kew for observing meteorological, electrical and magnetic phenomena. The Observatory was in the charge of Francis Ronalds, and Wheatstone had responsibilities in connection with some of the instruments during the period 1843 to 1846. According to the geologist Sir Roderick Murchison the establishment of the observatory was largely due to 'Professor Wheatstone's energy and ability'.

Telemetry

In connection with the Kew Observatory Wheatstone designed a number of meteorological instruments which could record their observations or transmit them over telegraph wires. The first was his 'Telegraph Thermometer', described in a report to the British Association in 1843. In this device a clockwork mechanism lowered a platinum wire into the mercury in a thermometer tube and then raised it again over a six-minute cycle. The instrument was intended for carrying up in a balloon and was connected to the ground by two fine silk-covered copper wires, one connected to the platinum wire and the other to the mercury. The points on the thermometer scale corresponding to the upper and lower limits of travel of the platinum wire were known, so by timing the periods when the circuit was open and closed it was possible to deduce the reading of the thermometer. Wheatstone pointed out that it was not necessary for the clock on the ground to keep exactly in time with the clock in the balloon.

In the following year, 1844, Wheatstone described more elaborate instruments which produced a printed record of their observations. There were three instruments, a mercury barometer, an ordinary dry-bulb thermometer and a wet-bulb thermometer. The recording equipment had a wire which was raised and lowered by clockwork in the tubes of the barometer and the thermometers, and made contact with the mercury during the lower part of its travel. The wire was raised in five minutes and lowered in one minute. As it moved, the same mechanism drove two typewheels which together gave 150 different readings. When the wire left the mercury a circuit was broken and a hammer mechanism, similar to that in Wheatstone's first printing telegraph, printed the number indicated by the typewheels on a paper strip. The mechanism could run for a week without attention and record an observation every six minutes.

There is a drawing of Wheatstone's recording meteorological equipment in a translation by C.V. Walker of Kaemtz's *Meteorology*, first published in Germany. Wheatstone possessed a copy of the book in French, published in 1844. Walker's translation appeared in 1845. There is a letter from Wheatstone to Ronalds in which he wrote 'Mr. Walker is publishing a translation of Kaemtz's Meteorology, and wishes to insert in it one month's register of the electrical etc observations made . . .'. Walker was not able to do so, for the apparatus was not ready in time. In the book Walker added at the end of his description, '. . . this register is for the Kew Observatory, where, we hope, very soon, to see it placed'. The instrument is now in the Science Museum.

Wheatstone might well be called the father of telemetry.

Typewriters

During the 1850s Wheatstone made three typewriters which are now in the Science Museum. Wheatstone did not patent his typewriters, nor did he publish on the subject. Presumably, therefore, he regarded them as unsuccessful. The mechanism of all three is similar and is a development of his first printing telegraph (see Figure 13.5). The letters are mounted on springy arms but in a

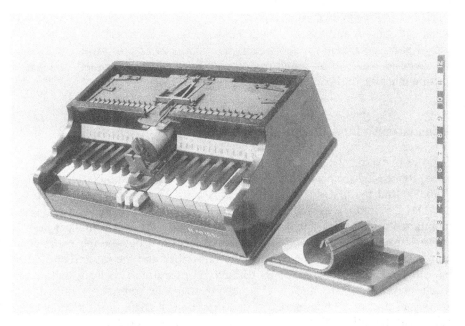

Figure 13.5   Two typewriters by Wheatstone (one partly dismantled).

line, not in a circle as in the telegraph. Operation of the keys moves the assembly of springy arms so that the appropriate character is in position, and a hammer then strikes it. The paper, which is wrapped round a cylinder, is moved forward as the key returns. The type is inked by a pad which moves underneath it before the hammer strikes. Two of the machines have a piano-type keyboard; the other is operated by a bank of 26 round knobs. Two of the machines also have a change of case, so that upper and lower case letters may be typed by the same keys. The first typewriter capable of practical work was the Burt Typographer of 1829, but the commercial development of typewriters did not begin until 1873 when C.L. Sholes and Carlos Glidden entered into a contract with E. Remington & Sons, whose previous products were small arms and sewing machines. Wheatstone was one of a number of people who had attempted to make a writing machine in the intervening period.

Wheatstone's later telegraphs

In order to exploit the ABC Telegraph patents Wheatstone established the Universal Private Telegraph Company in June 1861. It had an authorized capital of £190,000 and the first directors included Edward Franklin, William Fairbairn and David Salomons, as well as Wheatstone. The company would construct and maintain a private telegraph wire between two places for £4 per mile per year. It differed from all other telegraph companies in that it put the means of communication directly into the hands of the subscriber, and the subscriber did not have to send and receive messages through an operator. This feature attracted much private custom, and it also meant that the company required few employees and no public offices, with a consequent saving in overheads compared with the other telegraph companies. The company proved profitable, and when the telegraphs were taken over by the government in 1869 the compensation paid to the Universal Company's shareholders was £184,241 which was the estimated value of 20 years' profits. In addition Wheatstone received £9,200 for his ABC Telegraph patent rights. At nationalization the Universal Private Telegraph Company had over 2,500 miles of wire and 1,700 instruments in use. The telephone eventually replaced the private telegraph, although the last instrument on a Post Office line was not withdrawn from service until 1950.

Wheatstone also had his own manufacturing establishment, the British Telegraph Manufactory, which became a limited company in January 1874. It was wound up voluntarily in 1882. Robert Sabine, who married Wheatstone's daughter Catharine, was manager of the Manufactory and a shareholder in it. The Manufactory was first established in about 1858 when Wheatstone patented his first practical ABC telegraphs. Among its early orders was one for the ABC telegraphs which were installed between police stations in the City of London in September 1860 at a cost of £600.

For the last two decades of his life Wheatstone had the services of J.M.A. Stroh (1828–1914). Wheatstone's last four patents were obtained jointly with

Stroh, including one for musical instruments. The other three are all concerned with further improvements in telegraphs, and their main interest is that they show that Wheatstone continued to refine his instruments up until the end of his life.

Stroh was born in Frankfurt-am-Main and apprenticed to a clock and watch-maker there. He visited London in 1851 for the Great Exhibition, and was so impressed with what he saw of the state of science in England that he decided to stay, and became a naturalized British subject. In the mid 1850s he was introduced to Wheatstone, and thereafter worked closely with him on the 'ABC' and 'Automatic' telegraphs. Stroh became a Member of the Society of Telegraph Engineers in February 1875, and was twice a member of its Council. Like Wheatstone, Stroh was a man of very wide scientific interests. He worked on photography, horology, the phonograph and free-reed musical instruments, as well as on electricity and telegraphy.

The basic needle telegraphs used on all the early telegraph lines had the great disadvantage that someone had to watch the receiving instrument continuously while a message was being received. Usually the person watching would dictate the message to a colleague who would write it down. In practice, therefore, two people had to be on duty at each telegraph station whenever the system was operating. The automatic telegraph had two great advantages: a message could be received when no operator was in attendance and, with automatic transmission as well as automatic reception, a message could be sent much faster than by purely manual operation. Speed was important because it enabled more messages to be sent over one circuit, and the provision of the telegraph wires was the major part of the capital cost of a telegraph system. Several automatic systems were tried by the telegraph companies. All used punched tape which was prepared by clerks at the sending station. Several clerks could prepare message tapes simultaneously. The tapes would then be fed into a single transmitting instrument and the messages received on a single instrument at the receiving station.

Bain's chemical printing telegraph was in use from the early 1850s. The receiver used a paper tape soaked in a chemical, such as potassium iodide, which was colourless but could be decomposed electrolytically to give a coloured product. The tape was moved between metal contacts connected to the telegraph line so that the message appeared as a series of coloured marks in Morse or similar code. The Electric Telegraph Company experimented with several automatic telegraphs during the 1860s, including from 1867 Wheatstone's system which was widely adopted, being the only automatic telegraph that could exceed 40 words per minute.

On 2 June 1858 Wheatstone obtained two patents. One was for the improved ABC telegraph described above, the other was for his new automatic telegraph system, which he continued to develop for the rest of his life. The feature which really distinguishes Wheatstone's automatic telegraph from Bain's and all the others is the method of 'reading' the punched tape in the transmitting instrument to produce electrical signals. Bain's tape reader simply pulled the tape across a metal surface, and metal springs pressing on the tape made contact through the holes. Wheatstone's tape reader is quite different.

In Wheatstone's automatic system (see Figure 13.6) the tape had three lines of holes, the central line for driving and guiding, the outer lines punched, or not, according to a code. For a Morse dot there are a pair of holes on both sides of

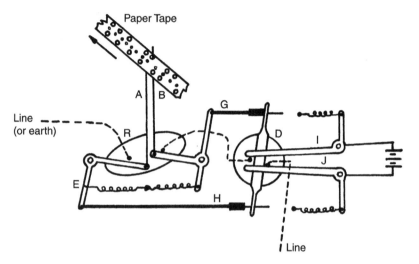

Figure 13.6   Wheatstone's automatic telegraph transmitter.
In this instrument two feeler rods A, B sense the presence or absence of holes in a paper tape and cause appropriate positive or negative pulses to be transmitted.

The elliptical rocker piece R is of insulating material and it carries two metal pins which can press on and make contact with the levers E, F. The rocker R is normally horizontal but at each tape hole position it is rocked first anti-clockwise (as shown) and then clockwise. The disc D is also of insulating material and carries two contact pins which make contact with the levers I, J when the disc D is turned a little in either direction. The disc D is driven by rods G, H which push (but cannot pull) the arms on the disc.

As shown the feeler rod B has risen because there is a hole in the tape at that point. Consequently the disc D has been turned clockwise and the telegraph lines are connected to the battery. (If there had been no hole at that point the feeler rod B would have been restrained by the paper and the circuit would be broken because the lever F would cease to make contact with its pin on the rocker R. Consequently no signal would be sent.)

The rocker then moves clockwise. As the code is a dash (two holes staggered) rod A is restrained and no further signal is sent at this point. If the code were a dot (two adjacent holes) the rod A would rise, the disc D would be turned anti-clockwise and a signal would be sent in the opposite direction to the first. The tape is then advanced to the next hole position and the sequence repeated.

The result is that for both dots and dashes the instrument sends a signal first in one direction then in the other, but the interval is longer for dashes.

the centre line at the same position; for a dash there are also two holes but they are staggered. The tape is drawn through the tape reader by the central line of holes, one hole-spacing at a time. At each position two light metal rods are pressed successively against the paper. The rods operate contacts whenever they pass through a hole, producing positive and negative pulses respectively. When a dot is read the two pulses occur in quick succession. When a dash is read their separation is increased.

The same specification includes a description of a tape punch with three keys. The central key is arranged to punch the central guide hole only, and advance the tape one space. The outer keys punch the signal holes as well as punching the central hole and advancing the tape.

Usually the automatic telegraph was used to send signals to a simple recorder such as the 'Morse inker'. This was operated by a polarized armature and would begin writing in response to a current in one direction and stop writing in response to an opposite current. It therefore printed dots and dashes on a moving paper tape. In the specification, however, Wheatstone described a 'translator' for converting the received signals into alphabetic characters using a special code but the same transmitting apparatus. The 'translator' had a typewheel with 30 characters which was rotated to the desired character and then made to print.

Strange though it may seem, Wheatstone published no account of the automatic telegraph in Britain, apart from the patent specification. He did, however, give an account of it to the Paris Academy of Sciences. The description published in *Comptes Rendus* differs in one respect from that in the specification: the French tape has central guide holes only where there are no signal holes. Possibly this represents an earlier stage of Wheatstone's thought and he made the line of guide holes continuous in the light of experience. During 1858 and 1859 Wheatstone was in correspondence with J.B. Dumas at the Paris Academy of Sciences, whom he visited to show his automatic telegraph in June 1858. Wheatstone was actively involved in the development of telegraphs in France, and felt that he had not been fairly treated by the French government when they took over the telegraph system.

After the initial invention of the telegraph, the most important step in its development during the nineteenth century was the introduction of the Wheatstone automatic system. In 1880 W.H. Preece, later Sir William Preece, Engineer-in-Chief of the Post Office, gave his assessment in a Friday Evening Discourse at the Royal Institution on the subject of Wheatstone's telegraphic achievements. He referred to the speed of Wheatstone's automatic apparatus:

When Wheatstone invented this apparatus it was only workable at an average rate of seventy or eighty words a minute, but scientific training, observation, thought, and care have resulted in improving upon this so much that, on making inquiry yesterday morning, I found that one of our Wheatstone instruments was actually working at the rate of 180 or 190 words a minute.

He then gave some figures for the number of Wheatstone instruments in use:

Everyone knows of the enormous development of the telegraphs. In 1870 the commercial part of the business was transferred to the Government, and at that time the business done in four weeks represented 554,000 messages. In the four weeks just expired it was 1,900,000. In the metropolis alone, while the number of messages of all sorts dealt with in four weeks in 1870 amounted to 130,000, in the four weeks just passed there were 726,000 . . .

But it is in the transmission of news where Wheatstone's telegraphic achievements have proved of such marvellous benefit. In 1871 there were distributed to the different papers copies of messages, some 2000 words long, others as short as 10 words, a total of 32,000. In 1879 they amounted to nearly 50,000. The number of words delivered in one week in 1871 was 3,598,000; in 1879 they amounted to nearly 6,000,000, which means 300 millions for the year, or 15,000,000 columns of the 'The Times'. There is not a town in the United Kingdom possessing a daily newspaper that is not in direct communication with London for news purposes, and by this means every man receives at his breakfast table the latest item of news, Parliamentary or general, just as readily as we do in London. And all this is done by the Telegraph Department with the Wheatstone apparatus. In 1870 there were only six wires used for special press purposes, now there are twenty-four. Besides the million words sent a day, there are newspapers in Glasgow, Dublin, and Edinburgh that rent wires for themselves, fitted up with different kinds of apparatus, by which they transmit all the debates of the Houses, &c.

In 1870 the number of Wheatstone alphabetical instruments was 1200, now 5000 are in use. There are now 151 circuits worked by the Wheatstone automatic apparatus, in 1870 there were only eight. This system has proved its superiority for the rapid dispatch of news, and, in time, will no doubt be adopted by all countries employing the telegraph. I have not the slightest hesitation in saying that our telegraphic apparatus (thanks to Wheatstone) is at the head of that of the world, and my own impression is that the time is not far distant when even America will take advantage of the inventions we are now using.

For the rest of his life Wheatstone's chief interest and occupation was the further refinement of his telegraphs. When he became ill and died in Paris in October 1875 he had gone there partly to attend meetings of the Academy and partly to persuade the French telegraph authorities to test his latest telegraphic inventions. In an address to the Academy of Sciences just after Wheatstone's death, M. Tresca spoke of Wheatstone's continued work on improving the telegraph, saying that during all the time he had been associated with the Academy:

Wheatstone ne s'est pas arrêté; sa préoccupation constante était d'améliorer la télégraphie et ses applications.
(Wheatstone never stopped; his perpetual concern was to improve the telegraph and its applications.)

Note

The sources for this chapter are Wheatstone's papers and other references included in the text.

PART 4
SIR CHARLES

Chapter 14

Wheatstone at home

Until the beginning of 1847 Wheatstone was a bachelor academic interested in advancing scientific knowledge (and his own scientific reputation) and devoting all the time he could to that end. Together with his brother he had run the family business at 20 Conduit Street for many years, and they both lived on the premises. He held an important position at the still new King's College. He also had other business interests, mainly developed from his scientific work. He was a competent businessman, but the business activities were a matter of necessity rather than choice. His favourite pursuit was always scientific research, sometimes arising from the business and sometimes from his various other interests.

Marriage and family

Suddenly, or so it seems, there was a complete change in Wheatstone's life – or perhaps our impression of his life up to this point is quite wrong. At the age of 45 he married, settled into a new home and raised a family. Charles and Emma were married on 12 February 1847 and their first child, Charles Pablo Wheatstone, was born on 2 May, less than three months later. The fact that the baby was conceived well before his parents' marriage can hardly have escaped the notice of those who knew them. It certainly surprised me, and I wasted some time looking for the birth date because I assumed it would be later. If it caused any surprise at the time, no trace now remains; nor is there anything to indicate why the name Pablo was chosen. Understandably in the circumstances, it was a very quiet wedding. *The Times* did not notice it, though it did record Wheatstone's attendance the next day at a Royal Society Conversazione held by the Marquis of Northampton. The marriage certificate records that Charles Wheatstone, Bachelor and Gentleman of St Peter Hammersmith, son of William Wheatstone, Gentleman, married Emma West, Spinster of Christchurch, daughter of John Hooke West deceased, on 12 February 1847 at Christchurch Marylebone. The marriage was 'by licence', so no banns would have been called. The ceremony was conducted by the Rector, the Revd R. Walpole, and the witnesses were Caroline Drew and H.P.L. Drew, the latter presumably the 'Mr. Drew, friend and medical attendant' at Wheatstone's funeral nearly 30 years later.

Emma was described by those who knew her as 'a lady of considerable personal attraction'. The only known portrait of her is the stereoscopic pair of photographs of Emma, Charles and the three elder children taken by Antoine Claudet (Figure 14.1). She was some 11 years younger than her husband. Her father had been a tradesman in Taunton. Her sister, Rachel Elizabeth, had married Thomas Heaviside, a wood engraver from Stockton-on-Tees; they moved to London in about 1849 and settled in Camden Town, where their younger son, the distinguished electrical physicist Oliver Heaviside, was born in 1850. The Wheatstones and Heavisides seem to have been on close terms, and Wheatstone took a great interest in his nephews. He encouraged Oliver to learn French, German and, perhaps surprisingly, Danish, and it may have been the uncle's influence which led Oliver to begin his working life in a telegraph company. The elder son, Charles Heaviside, born in 1846, was musical and became proficient on the concertina invented by his uncle.[1]

The Wheatstones acquired a house in Lower Mall, Hammersmith, and Charles Pablo was born there. Hammersmith generally, and the Mall in particular, was a pleasant place in which to live. The Duke of Sussex (George III's sixth son) had a house there, where his equerry, Captain Marryat, wrote his famous novels. In 1825 the Duke laid the foundation stone of Hammersmith Bridge, which was the first suspension bridge in London and was replaced by the present structure in 1887. Wheatstone had shares in the Hammersmith Bridge Company. Nearby was Chiswick House, the London home of the Dukes of Devonshire. The Mall itself runs beside the Thames upstream from the bridge and much of it, including Lower Mall nearest the bridge, remains little changed today. One guide book says of it: 'into the Mall are gathered more good eighteenth century houses than in any other part of Middlesex. All of them are elegant and well-preserved . . . On the other hand the so-called "Georgian mansions" at the Hammersmith end of the Mall are poor imitations of their well-proportioned neighbours'.[2] It was in one of these 'Georgian mansions', or 'pretty little houses looking over the water to the trees and pasture of the south', that Charles and Emma Wheatstone made their home. The houses were not numbered so it is not clear which was theirs. Several gentlemen lived in Lower Mall, according to the earliest *Post Office Suburban Directory*, published in 1860, along with a boat-builder, a beer retailer, a shoemaker and a cowkeeper. Further along the Mall, the Chiswick Press was producing 'some of the finest typography of mid-Victorian times'. All in all, Charles and Emma Wheatstone must have found Hammersmith an agreeable area, physically and socially, in which to raise a family. Charles could easily ride the five miles to King's College along the Great West Road, and the route was also served by omnibus from Hammersmith to the City.

Having married, theWheatstones enjoyed a settled family life. Their other

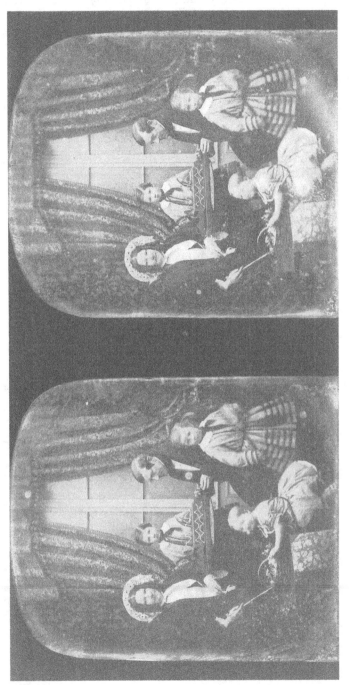

Figure 14.1   Charles and Emma Wheatstone with Charles Pablo, Arthur William Frederick and Florence Caroline. Stereoscopic daguerrotype by Antoine Claudet, 1852.

Source: By courtesy of the National Portrait Gallery, London.

children were also born at Hammersmith: Arthur William Frederick in 1848, Florence Caroline in 1850, Catharine Ada in 1853, and Angela in 1855. Catharine married Robert Sabine (1837–1884), who was for some years manager of Wheatstone's manufacturing works, which became the British Telegraph Manufactory. Florence and Angela were unmarried at the time of their father's death, though Florence later became Mrs Turle. Charles Pablo also married. There is a mystery surrounding Arthur. In his father's Will he received an annuity of £200, whereas his brother and sisters each received a lump sum of about £12,000, but there is no explanation for the different treatment, perhaps he was disabled and the annuity would provide care. Other bequests were to his surviving sister, who received an annuity of £100 and a house rent-free for the rest of her life, and to his niece Emma, the daughter of his other sister, who received an annuity of £50.

The Wheatstones remained at Hammersmith at least until 1859, but then moved to North London. After living briefly at 7 Chester Terrace they moved to 19 Park Crescent, a house which still stands on the southern edge of Regent's Park, and now bears a Blue Plaque recording that Wheatstone lived there. It was here that Emma died of kidney disease, at the age of 51, on 21 January 1865. Wheatstone remained at this address for the rest of his life.

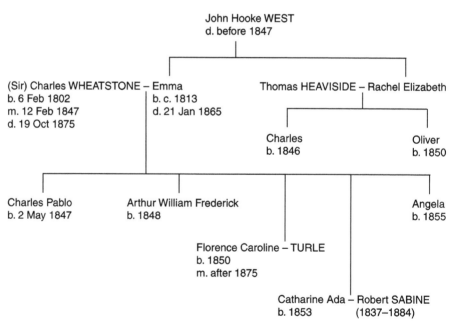

Figure 14.2   Wheatstone's family (see also Figure 2.1).

Private life

Leopold Martin, son of Wheatstone's friend, the painter John Martin, gave his impression of the house in Park Crescent in his *Reminiscences*:

The private residence of Sir Charles Wheatstone, in Park Crescent, Regent's Park, became one of the most interesting in London. Sir Charles had a perfectly arranged series of wires communicating with the electric department of the General Post Office, and it was his delight to startle a visitor or friend by sending a message to and receiving a reply from connexions or friends in distant parts of the country or abroad. His residence was at once one of the most scientific and the most charming in the metropolis, and the resort of all distinguished in art, science or literature.[3]

His circle of friends was by no means restricted to fellow scientists, and he was a welcome guest in many houses. The dinner parties at the actor Macready's and the visits to Dillwyn at Swansea and to the Nightingales have already been mentioned. Another friend was the artist and cartoonist George Cruickshank (1792–1878), who illustrated books for Dickens and other novelists of the time.

Wheatstone's views on contemporary issues are rarely apparent. We know nothing of his politics, though there is one Conservative party circular among his papers at the College, and of his religion we know only that he must have been a member of the Church of England to hold his post at King's College. We do know that he had strong views on the fashion for scientific men to become involved with spiritualism. He had spoken about it with the Hon. Lionel A. Tollemache who, with his wife, published a collection of their essays and poems under the title *Safe Studies*. (The volume, which went through several editions in the 1890s, was so called because it was quite safe to give the book to anyone; nothing in it could possibly give offence.) One of the essays was entitled 'Recollections of Sir Charles Wheatstone' and in it Tollemache recalls that:

Sir Charles told me that he had a curious conversation with the Emperor of the French [Napoleon III, reigned 1852–71] about Spiritualism. The Emperor gave an account of sundry marvels he had seen, and asked whether Wheatstone could explain them. Sir Charles admitted that he could not; but said that many feats of legerdemain seem inexplicable until they are explained.

Tollemache goes on to quote Wheatstone's opinion of scientific men who witness séances:

most of them are unwilling to suspect, and unable to detect, imposture . . . if any experts are to be present at those sad exhibitions, they ought to be professional conjurors. Not that he regarded all the reputed marvels as springing from imposture: many of them he ascribed to delusion. Both he and Mr. Babbage spoke of *Russian Scandal* as the most instructive of games; for it shows that a story, when passed on, will often gather bulk as rapidly as a snowball when rolled. On one occasion he heard that a friend of his own had pulled off the boot of a medium who was floating in the air. But his friend, being cross-questioned, said that he himself had not seen this feat performed, but that he had been told that other people had seen it!

Wheatstone's opinion is the more interesting when it is recalled that, at least in his younger days, he had with considerable success demonstrated scientific effects with displays of 'magic', like the Enchanted Lyre. His remark that scientific men are unwilling to suspect and unable to detect imposture bears a remarkable resemblance to sentiments expressed in the *New Scientist* in 1974 when that periodical was assessing the then much publicized feats of Uri Geller.

There is some evidence that he took an interest in phrenology, the study of the relation between the exact shape of the head and one's mental abilities. This was a fashionable subject in the mid-nineteenth century and one of the leading exponents was Dr Elliotson, a near neighbour of Wheatstone's in Conduit Street. In 1832 the *Lancet* published an article by a Mr Weisten on dreaming and sleepwalking. It has been suggested that 'Weisten' was in fact Wheatstone, though personally I do not think so.

Such knowledge as we have of Wheatstone's affairs derives in part from his habit of making his rough notes on any piece of paper that came to hand, including things sent to him, and many of these survive at King's College. His scrap paper included the backs of committee agendas, business correspondence (mainly circulars), 'begging' letters and advertising circulars. These throw light on some of his involvements. He was on numerous committees of the Royal Society, the British Association and the Royal Institution. He was concerned with the Royal Botanical Society from 1872 onwards. There are company papers relating to the Hammersmith Bridge Company, the West Chiverton Mine, the Mineral Hill Silver Mines Co Ltd, the Nevada Land and Mining Co Ltd and the Star and Garter Company, as well as various telegraph companies. A letter dated January 1870 from an estate agent reads 'Sir Charles, We can offer a most desirable tenant (the Bishop of Gloster [*sic*]) for a house in Park Crescent for the season months, if you should have any desire to let'. This was probably just a circular letter from a hopeful estate agent; there is no evidence that Wheatstone did let his house. There are invitations to subscribe to worthy causes, such as the Middlesex Hospital Home for Nursing and the St Augustine Club for Working Men. The occasional receipt shows that he contributed to some of these at least.

The Society of Telegraph Engineers, now the Institution of Electrical Engineers, was founded in 1871 and Wheatstone was a member from the beginning. In 1872 the Society invited him to speak on any subject he chose, though he did not do so. He never did address the Society, though he was a member of the Council and would probably have become President had he lived a little longer. At the first meeting of the Society after his death, the then President, Latimer Clark, gave a long address about the life and work of Wheatstone, an 'eminently scientific man'. Several members questioned Clark's historical accuracy in a few details, though not his assertion that:

a thousand years hence our successors will hear the name of Watt in connection with the steam engine, and of Stephenson in connection with the locomotive and railways, and they will also hear of Wheatstone in connection with the electric telegraph.

Wheatstone made many visits to Europe. Perhaps he was happier to be in a foreign gathering than an English one. His shyness was certainly a problem to him in England, but abroad his reluctance to address a public meeting would have been less embarrassing. He knew some German, Danish and Italian, and was fluent in French; indeed, he wrote some of his scientific papers in French. The most important of these was his paper on the automatic telegraph, which was never published in English.

The impression Wheatstone left on one who met him was recorded by the diarist Charles Greville (1794–1865). After Wheatstone had given him an account of the progress of the telegraph, Greville wrote:

There is a cheerfulness, an activity, an appearance of satisfaction in the conversation and demeanour of scientific men that conveys a lively notion of the pleasure they derive from their pursuits.[4]

Notes

1 Information about the Heaviside family and the relationship between the Wheatstones and the Heavisides is from the *Heaviside Centenary Volume* published by the Institution of Electrical Engineers in 1950.
2 Bruce Stephenson, *Middlesex*, 1972.
3 For Martin's *Reminiscences* see Notes to Chapter 7.
4 Greville's diary is quoted in Helen M. Fessenden, *Fessenden – Builder of Tomorrows*, New York, 1940.

Chapter 15

Later scientific work

In the 1850s Wheatstone was again working on the generation of electricity and he developed a new kind of magneto-electric machine – the induction generator. In an induction generator both the permanent magnets and the coils are static, but a piece of iron moves in such a way that the 'reluctance' of the magnetic circuit varies and therefore the magnetic field through the coils varies and an electromotive force is induced in them. Such machines are sometimes called 'reluctance machines'. Because the only moving part is a piece of iron and there are no sliding contacts, these machines are reliable and robust.

Wheatstone's first induction generator was made for the government's Ordnance Select Committee. This body was established in 1855 to advise on the ordnance problems of the navy and the army. It replaced the former Select Committee of Artillery Officers which had been a purely military body. The Duke of Newcastle, who as Secretary of State for War appointed the members of the new committee, insisted that some civilian experts should be appointed – much to the disgust of the army members! One of the civilian experts was Wheatstone. He was appointed to the committee for two years in the first instance, and the appointment was subsequently renewed for a further two years. After 1859 the committee reverted to being a purely military body, though Wheatstone completed the work he was then doing for them.

As might be expected, Wheatstone advised the committee on military applications of the telegraph, but in July 1857 the committee took up the subject of the electrical detonation of gunpowder. As part of their work for the Select Committee, Wheatstone and F.A. Abel, the Chemist to the War Department, carried out investigations on 'The Application of Electricity from Different Sources to the Explosion of Gunpowder'. In the course of their work they studied different forms of fuse and different sources of electricity, and there can be little doubt that it was mainly Abel who worked on the fuses and Wheatstone who dealt with the electricity. They studied three sources of electricity and carried out practical trials with each. These were 'Electro-magnetic induction . . . as produced from the apparatus known as Ruhmkorff's Induction Machine' (the Induction Coil);

Sir William Armstrong's hydro-electric machine; and the magneto. They did not experiment with voltaic batteries because, they said, numerous accounts of such experiments were already published.[1]

New generator designs

Wheatstone and Abel reported that for most purposes the best method of detonating gunpowder was to use a fuse designed for the purpose by Abel fired by a magneto designed for the purpose by Wheatstone. The one exception was for the simultaneous detonation of a large number of charges, and in that cased the recommended the use of a Leyden Jar charged by Sir William Armstrong's hydro-electric machine. This rather unlikely-sounding machine, devised by the founder of the Armstrong engineering company, generated an electrostatic charge by the friction of steam issuing from a jet. The output current was found to be proportional to the steam pressure, which could be up to 90 pounds per square inch. The machine took about half an hour to warm up, and had to be sheltered from draughts. After this report Wheatstone seems to have taken no further interest in Armstrong's machine, which lapsed into obscurity.

The machine which Wheatstone designed for the Ordnance Committee was an induction generator, for which he obtained a patent in 1858. In the induction generator the permanent magnets and the coils both remained stationary and pieces of soft iron moved in such a way as to vary the reluctance of the magnetic circuit. Because there was no commutator the machine was much more reliable than other contemporary generators. The output was alternating current but this was perfectly acceptable both for detonators, which utilize the heating effect of the current, and for the later designs of ABC telegraph (see Figure 15.1).

The following description, taken from the appendix to the report, was probably written by Wheatstone:

Description of Wheatstone's Magnetic Exploder

The magnetic apparatus employed in all the field experiments was designed especially for the purpose by Mr. Wheatstone.

It consisted of six small magnets, to the poles of which were fixed soft iron bars surrounded by coils of insulated wire. The coils of all the magnets were united together, so as to form, with the external conducting wire and the earth, a single circuit. An axis carried six soft iron armatures in succession before each of the coils. By this arrangement two advantages were gained; all the magnets simultaneously charged the wire, and produced the effect of a single magnet of more than six times the dimensions, and at the same time, six shocks or currents were generated during a single revolution of the axis, so that, when aided by a multiplying motion applied to the axis, a very rapid succession of powerful currents was produced. A single large magnet with a rotating armature could not be made to produce the same succession of currents without the application of considerable mechanical power. Another peculiarity of this apparatus was that the coils were stationary, and the soft iron armatures alone were in motion; by this deposition the circuit during the action of the machine was never broken. In the usual magneto-electric

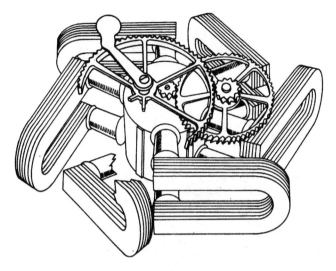

Figure 15.1   The arrangement of Wheatstone's earlier 'Magnetic Exploder'. The coils are fixed on the poles of the six horseshoe permanent magnets, and soft iron bars mounted on a rotating brass cylinder make and break the magnetic circuit as they pass by the coils.

machines with rotating armature the circuit is necessarily broken twice during every revolution, and this frequently gives rise to irregularities in the production of the currents. By the construction adopted, the currents can never fail to traverse the circuit.

The total weight of the instrument, enclosed in a case, as described in the Report, Part III, was 32 lbs. 11oz. It was enclosed for transport in a small packing case, weighing about 7 lbs.

Some further important improvements have recently been effected by Mr. Wheatstone in the magnetic exploder, whereby its size and weight have been very considerably diminished.

No machine of this design is now in existence, though photographs were taken in 1860 for Wheatstone's and Abel's report.

The Science Museum has an induction generator which is probably the improved machine referred to at the end of the above quotation from Wheatstone's and Abel's report (see Figures 15.2 and 15.3). In the 'Exploder' the magnetic flux through the coils was strengthened and weakened as the rotor bars approached and receded, but in the improved machine the flux is actually reversed: it changes from the peak value in one direction to the peak value in the other as the rotor turns. The electromotive force induced in each turn of a coil in this machine is therefore more than doubled for the same weight of permanent magnets. There is a single, stationary coil and there are six horseshoe permanent magnets. The upper poles of the magnets are alternately north and south. A cylindrical soft iron core rotates within the coil, and this core has three equally spaced projections at the top and at the bottom which pass close to the poles of the magnets as the core turns. As the core rotates the upper projections will, at a certain instant, all be adjacent to north poles and the lower ones adjacent to

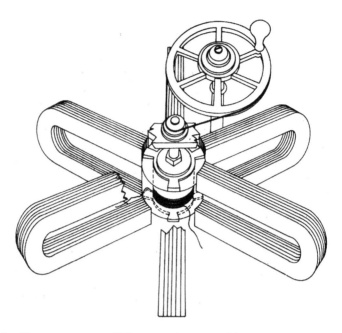

Figure 15.2   *The arrangement of Wheatstone's improved 'Magnetic Exploder'. The upper poles of the horseshoe magnets are alternately north and south. The single coil in the middle does not move. The moving piece is an iron cylinder with three horizontal projections above and three below. As the iron rotates a magnetic field is produced in the coil, first in one direction then in the other.*

south poles. One sixth of a revolution later the polarities, and the flux through the coil, are reversed.

The small size of the projections as compared with the cross-section of the magnets is evidence that Wheatstone had no concept of the need to design the magnetic circuit taking into account the reluctance of its constituent parts. The magnetic *flux* density in the projections is many times greater than that in the permanent magnets or the central cylinder. The projections could easily have been made of much larger cross-section if the magnetic circuit had been properly understood. In fairness to Wheatstone, however, it must be remembered that no machine designer considered the magnetic circuit properly until John Hopkinson in 1886.

Wheatstone seems to have played no further part in the development of generators for electrical detonation. In the later 1860s, however, Werner von Siemens developed a hand-driven self-excited generator with an interrupter mechanism for detonating mines.

In 1858 Wheatstone brought out a new design of ABC telegraph using an induction generator. The new instruments were a great improvement over the 1840 design and considerable numbers were installed by the Universal Private Telegraph Company during the 1860s. The sending instruments had a magneto

*Figure 15.3   Wheatstone's 'Magnetic Exploder' of about 1860 in the Science Museum.
One of the six field magnets has been removed to show the construction.
The only part which moves is a shaped iron cylinder in the middle of the
machine with three radial projections top and bottom.*

with a single fixed horseshoe magnet, coils mounted on soft iron extensions to
the poles, and a rotating soft iron strip which varied the magnetic flux in the
coils.

Waveforms

There is an experimental magneto of this pattern in the Science Museum (see
Figure 15.4) as well as several of the telegraph instruments. The experimental
machine is arranged to be driven by a falling weight, but its most interesting
feature is a special commutator. This has a thin contact segment which can be

Figure 15.4   *Wheatstone's experimental weight-driven magneto, showing the special*
contact arrangement which allows the output to be sampled at any one of
32 positions in a cycle. The coils are fixed to the poles of the
permanent magnet. Only the soft iron strip in front of the coils rotates.

turned relative to the rotor, so as to determine the instantaneous output at different points in the cycle. There were no oscilloscopes in Wheatstone's time, but by connecting a galvanometer through the special commutator arrangement, and observing the galvanometer readings when the magneto is run with the contact segment at different angles relative to the rotor, it is possible to build up a picture of the output waveform. Wheatstone could thus have observed the effect on the output waveform when the design of the machine was modified.

He would not have been seeking a sine-wave output. Indeed, it probably never occurred to him that an alternating current waveform might be a sine-wave, and in the days before transformers and before polyphase systems there would have been little point. He would probably have liked to produce an output waveform which was always at a peak value, either negative or positive – that is, a square wave. This would have been ideal for his later ABC telegraph, which required alternate positive and negative pulses to drive a step-by-step mechanism. Certainly he would have wanted a waveform whose negative-going and positive-going portions were similar.

The telegraph generator which Wheatstone developed after this study had four coils, two on each pole of a single horseshoe magnet. The four coils were placed at the corners of a square, and at any time the rotating soft iron armature was always approaching two coils and receding from two coils. Thus with the

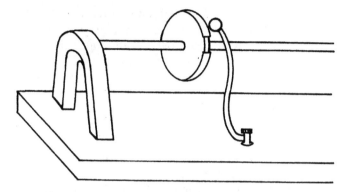

Figure 15.5   The contact arrangement on Wheatstone's experimental magneto. The circuit is completed for ½ of a revolution.

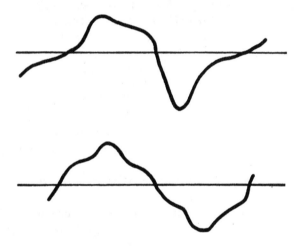

Figure 15.6   Output waveform from a magneto with two coils (above) and with four coils (below).

four-coil machine Wheatstone obtained an output waveform whose positive-going and negative-going halves are symmetrical, both being the sum of the 'approaching' and 'receding' waveforms of the two-coil machine, and the portion of the cycle during which the output is almost zero is much shorter. He did not get a square wave, but he is entitled to the credit for being the first to attempt to design a generator so as to obtain a desired output waveform.

There is no written account of Wheatstone's concern with waveforms, and this section is based solely on a study of the machines themselves. The experimental machine can still be used with a galvanometer in the way described to find the actual waveform. I made a number of measurements with it running at about 70 revolutions per second. The output was then a 140 Hertz alternating current. The electromotive force of the generator at that speed is about 20 volts, its internal resistance 550 ohms and its inductive reactance 2500 ohms. The contact segment on the commutator can be turned to any one of 32 distinct positions and the voltage measurements below were made with a lightly damped 500 ohm voltmeter connected to the generator through the special contact. The voltage readings obtained in this way are very small because the voltmeter is connected for only about $\frac{1}{32}$ of the time but, apart from the possible errors noted below, the reading is proportional to the instantaneous electromotive force of the generator at the particular point in the cycle. Positions 16 to 31 gave the same results as positions 0 to 15, since the rotor is symmetrical. Positions 0 and 16 are when the rotor lies along the line joining the poles. At positions 8 and 24 the rotor is at right angles to the line joining the poles.

A graph was plotted from the figures in Table 15.1 and compared with the waveform of the generator as shown by an oscilloscope. This showed that the method produces accurate results. Two potential sources of error were found if the generator ran much faster than 70 revolutions per second. These were monitored by connecting an oscilloscope across the voltmeter. First the wiper on the contact segment tended to bounce: this could be seen as 'chatter' on the oscilloscope and as a fall in the voltmeter reading as the generator accelerated much above seventy revolutions per second. This factor, coupled with the fact that it was difficult to make the machine run steadily at a much lower speed, was the reason for selecting 70 Hertz as the speed for the tests. The second source of error also worsened as the speed increased. When the wiper makes contact with the contact segment the current driving the voltmeter takes an appreciable time to build up; if the duration of contact is inadequate, the voltmeter will never reach the correct reading.

Despite its drawbacks, Wheatstone's method gives a good representation of the output waveform of a generator, and he seems to have been the only person to attempt this before the introduction of alternating current supplies.

Wheatstone's method of determining a voltage waveform was reinvented, apparently quite independently, by the French physicist J. Joubert in 1880 and was widely used in electrical laboratories under the name of the 'Joubert contact' until cathode ray oscilloscopes became commonplace in the later 1920s.

Table 15.1   Voltage shown at each rotor position

Position	Voltage	Position	Voltage
0	+0.01	8	0.00
1	+0.10	9	−0.06
2	+0.27	10	−0.18
3	+0.30	11	−0.18
4	+0.20	12	−0.16
5	+0.10	13	−0.14
6	+0.05	14	−0.08
7	+0.02	15	−0.06

Joubert does not mention Wheatstone in his paper and there is no drawing of his apparatus but the description leaves no doubt that it was a similar arrangement to Wheatstone's. Joubert's discussion is based on a circuit in which a Siemens' generator supplies a Jablochkoff's Candle with alternating current of a frequency of 100–200 Hz and the special contact connects for about $\frac{1}{20\,000}$ of a second in each cycle.[2]

The self-excited generator

The study of waveforms may have been an important factor in the development of Wheatstone's self-excited generator. By far the most important step in the evolution of the generator as we know it today was the practical introduction of the concept of self-excitation – that is, the idea that the magnetic field in the machine should be produced by an electromagnet energized by the output of the generator itself. The concept of self-excitation took a long time to develop, perhaps because it sounded rather implausible. The idea of using the output of a generator to produce the magnetism on which the generator depends sounds rather like lifting oneself up by the shoelaces!

Several people, including Wheatstone and the Siemens brothers, were working on the idea in 1866, but it was by no means a new idea then. Soren Hjorth, a Dane, obtained British patents in 1854 and 1855 for machines with permanent magnets and electromagnets. The electromagnets were energized by the current from the armature. With the benefit of hindsight we can see that Hjorth was working on the right lines, but his work seems to have come to nothing. His patent specifications are not easy to understand.

In the specification of the 1855 patent Hjorth wrote that in his magneto a series of coiled armatures 'in a wheel revolving at a slow motion are brought in succession between the poles of permanent magnets and the poles of a series of electromagnets', and the action is that 'the permanent magnets acting on the armatures brought in succession between their poles induce a current in the coils of the armatures, which current, after having been caused by the commutator to flow in one direction, passes round the electro-magnets, charging the same and acting on the armatures'. His machine probably failed because the permanent magnets and the electromagnets were not acting simultaneously on the armature, and when the armature conductors reached the electromagnets the energizing current had already ceased.

Wheatstone would have been aware of Hjorth's work, at least as soon as it appeared in the patent literature. The use of electromagnets instead of permanent magnets is specifically mentioned again in Wheatstone's 1858 patent. There is nothing to suggest that he tried to make a self-excited machine at that time, though the idea that the output of a machine might somehow be used to produce the necessary magnetism must have remained at the back of his mind, awaiting some fresh stimulus to bring it out (see Figure 15.7).

The stimulus came in April 1866 when a paper by Henry Wilde was read at the

Figure 15.7   Wheatstone's self-excited generator, made in 1866 and demonstrated to the Royal Society in 1867. The ropes and pulleys enable it to be driven by two men.

Royal Society, entitled 'Experimental Researches in Magnetism and Electricity'. It had two main sections. The first was on 'some new and paradoxical phenomena in electromagnetic induction'. He describes experiments which show that the lifting power of an electromagnet energized by a magneto-electric machine can be greater than the lifting power of the magnets of the magneto. He regarded this as a paradox and as an infringement in some way of the principle of the conservation of forces. It was in fact a commonly held view that the output of a magneto-electric machine was subject to two fundamental limits, determined by the mechanical force expended and the strength of the permanent magnets.

The second part of Wilde's paper described an arrangement which utilized the increased magnetic power to produce a more powerful generator than any previously known. He had two machines driven at the same speed. In some experiments the armatures of the two machines were identical; one machine had a permanent magnet and the other an electromagnet. The current generated in the armature of the permanent magnet machine was employed to energize the electromagnet of the second machine. The armature of the second machine gave an output eight times that of the simple magneto on its own.

Wilde's paper was read in April 1866. Successful self-excited generators were announced on 14 February 1867 when C.W. Siemens and Wheatstone presented similar papers to the Royal Society. When two men announce virtually the same discovery simultaneously, a natural question to ask is who really got there first. In this case there can be little doubt that their work was quite independent, although after publication they discussed their machines and experiments together. Other men were working on similar lines at the same time, but Siemens and Wheatstone share the honour of first announcing the discovery.

Both papers explain that an armature rotating between the poles of an electromagnet possessing some residual magnetism will generate electricity, and that if the armature is connected through a commutator to the winding of the electromagnet then the magnet is strengthened, so that more current is produced, so that the magnet is strengthened further, and so on. The two papers appear together in the *Proceedings of the Royal Society*, and although their subject matter is almost identical the two papers reflect the differing interests of their authors.

Siemens' paper, which was largely inspired by his brother Werner in Berlin, reads like the work of a practical engineer. The title was 'On the Conversion of Dynamical into Electrical Force without the aid of Permanent Magnetism', and he ended with the conclusion that 'It is thus possible to produce mechanically the most powerful electrical . . . effects without the aid of steel magnets, which . . . are open to the practical objection of losing their permanent magnetism in use.'

Wheatstone's paper, on the other hand, is written as a more scientific work. He chose a more learned-sounding title, 'On the Augmentation of the Power of a Magnet by the reaction thereon of Current induced by the Magnet itself', and he sets out to show that an electromagnet possessing some residual magnetism may be made into a powerful magnet by currents originated by the action of the magnet itself. His machine had a stand equipped with ropes, pulleys and handles for two men to turn – probably two of his students demonstrated it.

The self-excited generator is fitted with the special contact arrangement described above which would have enabled Wheatstone to determine the wave-form of the electromotive force induced in the armature. No record survives of any such trials he may have made, and it is no longer possible to run the machine. Information about the waveform could have been of interest to him for two reasons. First, Wheatstone may have suspected, as does the present writer, that Hjorth's self-excited machine failed because the field coils did not receive current from the armature at the exact time the armature conductors were underneath those coils. He would therefore have wanted to know when the electromotive force in the armature was at its peak value. Second, the arrangement of the magnetic circuit in the machine is different from that in any of his earlier machines. The pole pieces have a much more modern appearance and each pole piece extends nearly half way round the armature. Wheatstone may have intended his armature conductors to pass through a more uniform magnetic field than in his earlier machines, so that a more uniform and less peaky output was obtained. The special contact arrangement would have enabled him to determine the output waveform and see the extent of his success.

One piece of research on this machine did appear in the paper. Wheatstone discovered that if a shunt resistance was placed across the field winding then the output current was increased. After the Royal Society meeting Wheatstone wrote to Siemens suggesting that he should try shunting the field winding of his machine. Siemens tried it and told Wheatstone that it was beneficial. The explanation Wheatstone offered in his paper for this phenomenon was that although the shunt weakened the magnetic field, which would tend to weaken the current, it also reduced the circuit resistance, which would tend to strengthen the current, and with suitable values the latter effect was stronger so that the output current was increased. Subsequently Varley showed that Wheatstone was wrong in his theory, and he explained that the presence of the shunt resistance enabled a circulating current to flow in the field winding so that the field did not have to become demagnetized when the circuit was broken at commutation.

In his paper Wheatstone gave some information about the output of his machine. This was before the days of Volts, Amps and Watts, and there is some difficulty in converting Wheatstone's data into modern units. Without the shunt the machine was powerful enough to bring to red heat a piece of platinum wire 10 centimetres long and 0.017 centimetres in diameter. With the shunt it could heat 18 centimetres of the same wire. We also have the information that this was hard work for two men. If we knew how hot a piece of wire had to be on a February evening in 1867 to persuade the Fellows of the Royal Society that it was red hot, then we could estimate the efficiency of this generator. Siemens gave no data for the output of his machine, though he did say that it could run hot enough for the insulation to catch fire!

No record survives of anything further Wheatstone may have done on the generator in the remaining eight years of his life. Probably his experiments continued, but he made no further discoveries in this field.

Miscellaneous science

Although the greater part of Wheatstone's scientific work can be considered under the broad headings of Sound, Electricity, Telegraphy and the Stereoscope, he was a man of wide interests and he made a number of scientific contributions in other matters which caught his interest.

An example of Wheatstone noticing a scientific phenomenon and following it up with a series of experiments may be found in the 1844 *Reports of the British Association for the Advancement of Science*. The report in question is quite brief:

On a singular Effect of the Juxtaposition of certain Colours under particular circumstances Having had his attention drawn to the fact, that a carpet worked with a small pattern in green and red, when illuminated with gas-light, if viewed carelessly, produced an effect upon the eye as if all the parts of the pattern were in motion, he was led to have several patterns worked in various contrasted pairs of colours; and he found that in many of them the motion was perceptible, but in none so remarkably as those in red and green; it appeared also to be necessary that the illumination should be gas-light, as the effect did not appear to manifest itself in daylight, at least in diffused daylight. He accounted for it by the eye retaining its sensibility for various colours during various lengths of time.

Wheatstone was interested in the phenomenon of polarization of light at the time he wrote his first scientific paper, but his only research paper on the subject appeared almost 50 years later. In the meantime he had applied his interest in polarization to the solution of a practical problem: how to tell the time by the sun when the sun cannot be seen. In 1848 he described to the British Association several forms of instrument which he called the Polar Clock or Dial, which makes use of the fact that light from the sky is polarized and the plane of polarization depends on the position of the sun, and rotates once every 24 hours. Two polar clocks of designs described by Wheatstone and made by W.H. Darker, an instrument maker who worked from 1844 to 1866, are now in the Science Museum (see Figure 15.8). They both contain pieces of a polarizing material, selenite, which is used to observe the plane of polarization of light from the sky when looking north.

According to Wheatstone's description, one polar clock can be used to find the time to within one fifth of an hour (12 minutes), and the second form will indicate the nearest half hour. I could not achieve anything like this accuracy in London, although it was certainly possible to estimate the time using these instruments. Perhaps Wheatstone tried them out well away from the dirty city air which probably reduces the degree of polarization of the light from the sky. The polar clock was used on some polar expeditions during the arctic winter, though Wheatstone did not envisage that application in his original account of it.

In 1835 Wheatstone submitted a paper to the British Association 'On the Prismatic Decomposition of Electric Light', in which he showed that the light

Figure 15.8   Apparatus used by Wheatstone in studying polarized light.

from an electric spark may be analysed by a prism to give colour lines which are characteristic of the metals between which the spark is produced. Wheatstone deduced from that the important theoretical conclusion that 'electric light', by which he meant the light of an electric spark, resulted from the volatilization of the material of the conductor, and was not a visible manifestation of the electricity itself. This marked the beginning of the subject of spectrum analysis which has proved so important in the history of chemistry. Wheatstone did not publish anything else on the subject though his unpublished papers at King's College show that he continued to work on it in later years.

In 1855 he acquired a specimen of the metal aluminium, which had only recently been isolated, and reported to the Royal Society on its position in the Voltaic series.

In a delightful note in the *Proceedings of the Royal Society* in 1870 Wheatstone drew attention to 'a cause of error in Electroscopic Experiments'. He intended to write something on static electricity though this was one of the projects left unfinished at his death. In the course of a series of observations with an electrometer he was troubled by unexplained charges which appeared on his test objects after he had carefully dried them by holding them in front of the fire. He thought at first that the objects had been electrified in some way by the heat or by differences in the state of the air in different parts of the room, but by carefully planned observations he found that that was not the case. Eventually he found that the cause was the generation of static charges by the friction between his shoes and the carpet in a dry room. Furthermore he was wearing cotton socks which allowed the charge to pass from the shoes into his body. Woollen socks, he found, would insulate the charge. In these days of well-heated buildings and man-made fibres for carpets and clothing, we are used to the phenomenon of a charge building up on the body to give a shock on touching a metal object or another person. In Wheatstone's day it was a novelty, though he managed to find one account of it in print. Professor Loomis of New York had reported it at a meeting in Dublin in 1857, but the report, according to Wheatstone, was received 'with great incredulity' at the time.

In 1851 the French scientist J-B-L. Foucault (1819–1868) first gave his spectacular demonstration of the rotation of the earth with a long, heavy pendulum suspended in the Pantheon in Paris. He showed that the pendulum would continue to oscillate in a plane fixed in space and that therefore its plane of oscillation appeared to move as the earth turned in its daily rotation. Foucault stated that his pendulum demonstration was an example of the more general phenomenon that if an object is vibrating then the plane of the vibrations will not rotate if the point at which the object is fixed is rotated. He first observed this with a steel rod fixed in a lathe: when the rod was made to vibrate in a horizontal plane the vibrations continued horizontally if the lathe was turned.

Foucault's work seems to have caught Wheatstone's interest. Later the same year, 1851, he submitted a paper arising from it to the Royal Society. The mathematical analysis of Foucault's pendulum (except in the special case of a pendulum placed at the north or south pole) is difficult, as is the experiment, because the pendulum has to be released exactly in a straight line and because its motion might be affected by draughts. Wheatstone observed that as a result many people 'doubt either the reality of the phenomenon or the satisfactoriness of the explanation'. He therefore devised an experiment to demonstrate the general phenomenon in which the vibrating object was a spring (rather than a pendulum) and the apparatus was mounted on a wheel which could be turned rather than depending on the rotation of the earth. A wooden semicircular arch is fixed on a horizontal wheel and the spring is fixed between the centre of the arch and a slider on the arch. The spring is set vibrating by plucking and its plane of vibration can be observed as the wheel is turned.[3]

In 1852 M. Fessel of Cologne described the gyroscope which provided another mechanical proof of the rotation of the earth. In a note to the Royal

Society Wheatstone described a more elaborate form of gyroscope than Fessel's. It had the spinning disc at the end of an axis, that was itself free to move about both a horizontal and a vertical axis, and the disc was counterbalanced by a weight. He described the precession of the gyroscope in response to various applied forces including magnetic forces.

In 1853 the Royal Society published a paper by Wheatstone 'On the Formation of Powers from Arithmetical Progressions', in which he gives a method of calculating any power of a number from an arithmetical series. He then demonstrates that the formula is valid for a number of specific cases, but he does not give a generalized proof. Wheatstone usually avoided mathematics if he could, and there is no difficult mathematics in any of his work. This paper is based on simple arithmetical calculations. For example, he states that every cube n^3 is the sum of an arithmetic progression of n terms, the first being n and the difference $2n$. Hence:

$$
\begin{array}{llll}
\text{If} & n = 1 & 1 & = 1^3 \\
& n = 2 & 2 + 6 & = 2^3 \\
& n = 3 & 3 + 9 + 15 & = 3^3 \\
& n = 4 & 4 + 12 + 20 + 28 & = 4^3
\end{array}
$$

There is nothing to indicate how Wheatstone became interested in this, or why he published this odd mathematical paper.

Adviser to government

On a number of occasions Wheatstone's talents were made available to the government and public bodies.[4] Mention has already been made of his service on the Committee of Inquiry into the Construction of Submarine Telegraph Cables and his work for the Select Committee on Ordnance.

He seems to have taken some interest in questions of public health and sanitation. After the cholera epidemic of 1831–1832 in which over 50,000 people died, there was widespread concern over matters of public health. John Martin proposed a scheme for building large sewers alongside the Thames under new embankments to convey London's sewage away from the town without fouling the Thames. A committee, whose members included Wheatstone and Faraday, was established to develop the idea, but nothing came of it at that time.

In 1842 Edwin Chadwick published *The Sanitary Condition of the Labouring Population of Great Britain* and this led to the establishment of the Health of Towns Commission. Medical and engineering experts were on this body, and they reported in 1844 and 1845. It became appreciated that poor living conditions, polluted water supplies, and lack of adequate drainage were major factors in the spread of disease and general ill health. At length in 1848 Parliament passed the Public Health Act and created the General Board of Health which was empowered to make regulations dealing with water supply and sewerage and to investigate anything injurious to health. The Board was established for five years in the first instance, but the Act was extended annually until 1858 when other government departments took over its functions.

In 1856 the Board invited a committee of five, Lyon Playfair, W. Fairbairn, J. Simon, L. Glaisher and Wheatstone, to investigate 'the best practicable methods of warming and ventilating Dwelling Houses and Barracks'. The following year the inquiry was extended to include schoolrooms. Regrettably there is nothing to tell us what Wheatstone contributed to these inquiries.

Wheatstone was also consulted about the design of Big Ben. On 16 October 1834 the old Palace of Westminster was destroyed by a fire which left only Westminster Hall intact. A public competition was held for the design of new buildings, and this was won by Charles Barry who was later knighted for his work. The new buildings were erected between 1840 and 1857 and survive to the present day except for the House of Commons which was bombed in 1941 and later rebuilt. Wheatstone was consulted at an early stage in the planning of the new buildings, though it may have been an informal and unofficial consultation. Among Faraday's correspondence is a letter of November 1838 from H.T. de la Beche saying that he was looking for stone suitable for building the new Houses of Parliament and had had a conversation at the Athenaeum with Wheatstone about testing stone to measure moisture absorption and its ability to withstand cold and pressure. It seems that de la Beche had written to Wheatstone but had received no reply two weeks later; he hoped that Faraday would remind Wheatstone about it.

Barry's design included the clock tower at the eastern end of the building, adjacent to Westminster Bridge. The clock itself and the bells were designed by Edmund Beckett Denison (1816–1905), later Baron Grimthorpe, a wealthy barrister and amateur architect. He was President of the Horological Institute for 37 years, and wrote *A Rudimentary Treatise on Clocks, Watches and Bells*. It has been said that the title of that book was a rare piece of modesty on Grimthorpe's part, and that he was one of the rudest and most crotchety of men.

In 1855 the Office of Works – the government department responsible for the building – wrote to Wheatstone seeking his views on the bells for the proposed clock tower. The letter, signed by an Assistant Secretary of the Office of Works, reads:

I am directed by the Commissioners of Her Majesty's Works, &c. to acquaint you that they are about to make arrangements for the supply of the Chimes required for the Great Clock of the new Houses of Parliament, and that they would feel obliged, should this be a subject to which your attention has been directed, if you would favour them with your advice in regard to the size, weight, and shape which should be adopted in the construction of these Bells.

Wheatstone was asked to confer with Barry if he was willing to assist in the matter. Presumably he was willing for the same Assistant Secretary wrote to Wheatstone again:

The Commissioners of Her Majesty's Works, &c. have had before them your letter of the

11th Inst. on the subject of the proposed Chimes for the Great Clock at the new Houses of Parliament for the construction of which you state no known bell founder in England can be relied on; – And I am directed by them to inform you that they approve of your suggestion that during the visit of yourself and Sir C. Barry to Paris as Jurors of the Industrial Exhibition you should collect information respecting the most esteemed Chimes in France and Belgium and whether there are in either of those countries makers acquainted with the conditions of the Art, or who have applied the modern discoveries of Science to the Improvement of Bells or to efficient substitutes for them. And the Board request that you will proceed in accordance with that suggestion.

The two letters quoted above are in King's College Library together with a third, written in August 1855, informing Wheatstone that Sir Benjamin Hall (the Commissioner of Works) was about to seek tenders for the casting of the bells and that he wanted Wheatstone to act with Mr Denison and Mr Taylor as 'referees regarding the Chimes'.

Wheatstone was probably an obvious person to be consulted about the bells, and he may already have known Denison. Denison was one of the first people to work out mathematically the relationship between the shape and weight of a bell and its musical note. This work would certainly have been of great interest to Wheatstone after his own study of the mathematics of various musical devices, and it could have brought the two men together. Despite Wheatstone's statement that no bell founder in England could be relied on to cast the bells, they were cast in England, presumably on the advice of Wheatstone, Denison and Taylor. Later events seem to have justified Wheatstone's original view. The hour bell – which even before it was hung was given the nickname 'Big Ben' after Sir Benjamin Hall – was cast in 1856 by the firm of Warner's at Norton, near Stockton-on-Tees. The 16-ton bell was taken by rail to West Hartlepool, then by boat to London. Finally it was dragged to Westminster by 16 horses. One thing Denison did not know, despite his mathematical study of bells, was how heavy the hammer should be. He had the bell hung on a temporary frame at the foot of the tower, experimented with hammers of different weight – and cracked the bell. Later, on 18 February 1858, the bell was broken up by having an iron ball weighing over a ton dropped on it. The *Illustrated London News* found the whole matter very funny and reported 'The Savants, Rev Taylor and E. B. Denison came as mourners of the bell . . .', but there is no mention of Wheatstone on that occasion although the paper had named all three men in earlier reports.

The broken pieces of the first Big Ben were taken to George Mears' bell foundry in Whitechapel, where the present bell was cast in 1858. It weighs 13½ tons. Denison was still involved, despite the failure of the first bell, so presumably Wheatstone and Taylor were also. The present bell is inscribed 'This bell was cast by George Mears of Whitechapel for the clock of the House of Parliament, under the direction of Edmund Beckett Denison QC in the 21st year of the reign of Queen Victoria, in the year of our Lord MDCCCLVIII'. After two months' service the new bell was found to be cracked. Mears said it was because the hammer was much heavier than he had specified. Denison said the bell had cracked when it was cast and Mears had concealed the crack. Eventually the bell

Figure 15.9   *Wheatstone was interested in everything. This scene shows him in the audience in the lecture theatre at the Royal Institution. The bronze panel by W.B. Fagan shows Michael Faraday lecturing and, from right to left in the front row, Tyndall, Huxley, Wheatstone (leaning forward intently), Crookes, Darwin, Daniell and Frankland.*

     Here in 1880 W.H. Preece said, 'Wheatstone's familiar form was very well known to the old habitués of this theatre. Whenever either of his favourite subjects, light, sound, or electricity, was under discussion, his little, active, nervous and intelligent form was present, eagerly listening to the lecturer.'

was turned slightly and a lighter hammer fitted, and it has remained in use ever since.

Notes

1   For an account of the Ordnance Select Committee see the *Official Guide to the Public Record Office*, the minutes of the Committee in the Public Record Office, and O.F.C. Hogg, *The Royal Arsenal – its background, origin and subsequent history*, 1963.

2 Joubert's paper was 'Sur les courants alternatifs et la force électromotrice de l'arc électrique' published in *Journal de Physique* 1880 and also in *Comptes Rendus* 1880. I am grateful to the late Professor James Greig of King's College London for drawing my attention to Joubert's work. Professor Greig read my account of the Wheatstone contact method and told me 'That's how we did it when I was young. We called it the Joubert contact'.

3 Wheatstone's work on Foucault's pendulum and the gyroscope is described in *The Earth and its Mechanism*, also published under the title *The Rotation of the Earth*, by Henry Worms, 1862. Worms dedicated his book 'to Professor Wheatstone, with feelings of the deepest respect, and in admiration of that fertile genius which has produced so many useful inventions in science and art'.

4 Letters to Wheatstone from the General Board of Health and the Commissioners of Works are in King's College Library.

Chapter 16

The public figure

Wheatstone received many honours. In King's College there are diplomas and medals relating to more than 30 awards and distinctions. He was elected a Fellow of the Royal Society in 1836 and later received its highest award, the Copley Medal. He became a Chevalier of the Legion of Honour in 1855 and a Foreign Associate of the French Academy of Sciences in 1873. He was honoured by governments, universities and learned societies in England, Scotland and Ireland, European countries, the USA and Brazil. He was knighted by Queen Victoria on 30 January 1868.

Some people felt he should have received a knighthood sooner. After the completion of the first successful transatlantic telegraph cable several of those immediately concerned were given knighthoods or other honours, though it could reasonably be argued that they had little to do with the transatlantic telegraph. Both Cooke and Wheatstone were omitted from the list. Most of these awards were first announced during a celebration banquet in Liverpool on 1 October 1866. Wheatstone was not present, though he did attend the banquet given by the Lord Mayor of London at the Mansion House on 30 October. Since no award was bestowed on Wheatstone at this time, *The Times* embarked upon a campaign to obtain an honour for him. On 9 October 1866 it published a letter signed 'J.D.', presumably John Dillwyn Llewelyn, praising the various honours bestowed and continuing:

But how happens it that in the midst of this jubilee and triumph the philosopher who invented telegraphic communication is left in the shade? . . . I cannot help thinking that . . . the man of science and genius, who elaborated his own idea and manifested its practicability, should not be forgotten in the dispensation of honours and dignities.

A leader next day takes up the theme in fulsome prose:

A correspondent has had the courage to express what many feel, but which they think will be set down as the old old remonstrance against the inequality of human rewards . . . Columbus led the way to the New World, but a humble imitator gave it his name . . . We see this failure of justice repeated in the story of the invention which has united the two worlds . . . If any name may be set at the head of the noble list in this new roll of honour it is that of WHEATSTONE . . .

The writer continues in similar vein for 140 lines!

After reading J.D.'s letter, Wheatstone wrote to Llewelyn and his brother asking what they remembered of the submarine cable experiments at Swansea 22 years earlier. Presumably Wheatstone minded being overlooked and was gathering evidence to show that he had pioneered telegraphs under water as well as on land.

A week later, on 15 October, *The Times* carried another letter on the same theme from Wheatstone's friend George Cruickshank, though Cruickshank's appeal is spoilt by his obvious lack of technical understanding of Wheatstone's work. This was followed by another leader written in similar style to the previous one but even more pointed.

The Times was then silent upon the subject for 15 months until January 1868 when it reported:

It is understood to be the intention of the Government to confer a title on Professor Wheatstone in consideration of his great scientific attainments and of his valuable inventions.

An official announcement followed shortly afterwards, and on 28 January *The Times* published a summary of Wheatstone's career, stressing his work for the government and in particular the fact that he had been associated with the Ordnance Select Committee at Woolwich during the Crimean War. The title, Knight Bachelor, was bestowed by the Queen at Osborne on Thursday 30 January 1868. No other awards or honours were given on that occasion. W.F. Cooke received his knighthood the following year.

Wheatstone never retired, but continued working to within a few days of his death at the age of 73. It was while attending meetings of the Academy of Science in Paris that he became ill with bronchitis and he died on 19 October 1875. Only one other illness is recorded in his whole life, in December 1843, when he was 'seriously indisposed' for a couple of weeks. A French writer, M.E. Mercadier, has left us a brief impression of Wheatstone as seen by foreign eyes. In an obituary he said that he had seen Wheatstone only a week before his death, in full health and chatting in his usual lively manner ('plein de santé, causant avec l'animation, la vivacité qui lui étaient habituelles').

Wheatstone's closest friend in Paris was the chemist J.B.A. Dumas (1800–1884), a leading figure in the Academy. Several English scientists were in Paris for the meetings, including Warren de la Rue. From their correspondence it is clear that the three men were on close terms, and that Dumas and de la Rue attended to their old friend in his last illness. Wheatstone's unmarried daughters, Florence and Angela, had accompanied him to Paris. Sabine was presumably in London running the British Telegraph Manufactory. Two days before Wheatstone's death de la Rue wrote to Dumas about a laboratory visit and added that he was very concerned about the health of his friend. He thought the daughters ought to send for their brother-in-law but did not want to worry them unduly,

Figure 16.1 Sir Charles Wheatstone in 1868. Drawn in chalk by Samuel Laurence.

Source: By courtesy of the National Portrait Gallery, London.

and he wanted to speak to Dumas about it. The previous day, 16 October, Wheatstone had made his will, and Warren de la Rue had witnessed the signature.

In the event Robert Sabine was summoned from London and arrived in time, probably with Catharine, thanks to the telegraph his father-in-law had done so much to establish. So Wheatstone died in a foreign land, but among family and friends.

Dumas spoke of his friend's death at a special meeting of the Academy two days later. He began:

Messieurs

Sir Charles Wheatstone s'éteignant mardi à deux heures aux milieux de sa famille en pleurs, accourue pour l'assister à ses derniers moments.

(Sir Charles Wheatstone died on Tuesday at 2 o'clock, surrounded by his weeping family who had been called to be with him in his last moments.)

Wheatstone's body was brought back to England and buried in Kensal Green Cemetery on 27 October 1875. We learn from *The Times* report that his son Charles Pablo and son-in-law Robert Sabine were there, but there is no mention of his other son, Arthur. 'A numerous assemblage of attached friends' were there, according to Brooks, including some with whom 'he had maintained an uninterrupted intimacy from his early manhood'. Cooke was there also, and a biographer recorded that 'any soreness of feeling which might once have existed had been outlived' ... 'Sir William F. Cooke was one of the most genuine mourners who assembled at the funeral of Sir Charles Wheatstone'.

Wheatstone's achievements eventually brought him public recognition and a fortune. He has sometimes been represented as a poor businessman, yet paradoxically he was also said to be very careful in financial matters. An article in *The Times* just after his death includes these conflicting assertions:

He was extremely precise and strict in pecuniary affairs, making all periodical payments, such as rent, wages, and the like, always at the exact time when they became due, and preserving his papers in the most methodical order. Notwithstanding this, he seemed incapable of keeping accounts, or of devoting his mind to what is ordinarily called business.

Wheatstone had no desire to be seen as a businessman or to devote time to business affairs. Nevertheless, he came from a family of successful small businessmen and was shrewd in such matters – hence his refusal to enter into partnership with Cooke on less than equal terms. Perhaps Cooke fostered the idea of Wheatstone as the scientist with no business sense and himself as the man of affairs who turned their invention into profit, but this view does not stand up to close examination. Cooke made even more money than Wheatstone from the telegraph, yet he died a poor man having lost all his fortune in other ventures. When Wheatstone died he left just over £70,000.

How did he acquire this sum? As we have seen, in 1845 he sold his interests in the early patents to Cooke for £30,000. He had invested an unknown sum in the partnership and his other scientific work had been restricted by lack of ready cash. At the end of the partnership his only financial assets were his share in the family business and £30,000. Until then he had been dependent on whatever he had inherited, and on the profits of the music business; his professorial income would have been very little. The £30,000 would have enabled him to establish a comfortable home of his own when he married in 1847.

The other main source of Wheatstone's fortune was the compensation he received when the telegraphs were nationalized a few years before his death. In exchange for certain patent rights he had received £17,000 worth of shares in the Universal Private Telegraph Company out of its total capital of £190,000. The shareholders received £184,000 from the government, making Wheatstone's share about £16,500. He also received £9,200 directly from the government for other ABC telegraph patents. The British Telegraph Manufactory, founded by Wheatstone, remained a private business until it was wound up voluntarily some years after his death, and it is more difficult to value his interest. In April 1875 the capital of the company was £16,325, and Wheatstone held 1,010 of the 3,000 shares. Some shares he had bought for cash, but 800 of them, worth about £4,400, had been received in exchange for patents. In all, therefore, Wheatstone received a little over £60,000 for his telegraphic inventions. His inheritance, the family business, and other interests had provided enough to keep him and his family and still leave another £10,000.

Some of the tributes paid at his death have already been quoted. On the 50th anniversary, 19 October 1925, a commemorative bronze plaque was placed on the outside of St Michael's Church tower in Gloucester, where it could be seen from the road. It cost £86, which had been raised by public subscription. The inscription read:

SIR CHARLES WHEATSTONE, D.C.L., LL.D., F.R.S.

Born at the Manor House, Barnwood, Gloucester, 6 February 1802. Pioneer of the electric telegraph and the first, with (Sir) Wm. F. Cooke, to render it available for the public transmission of messages 1837. Inventor of the stereoscope 1838. Conducted the first experiment in submarine telegraphy 1844. Contributed to our knowledge of acoustics and of spectrum analysis. Invented the rheostat, the polar clock, the automatic transmitter and receiver and the rotating mirror for determining the speed of electricity. Applied and improved the resistance balance of Christie, known as the 'Wheatstone Bridge'. Invented the self-exciting shunt-wound dynamo 1867. Knighted 1868. Died in Paris 19 October 1875. Buried at Kensal Green Cemetery, London.

Before the unveiling the Mayor of Gloucester opened a meeting in the Guildhall, at which there were speeches by the President of the Institution of Electrical Engineers (Mr R.A. Chatlock), the Director of the Science Museum (Col. H.G. Lyons), the Managing Director of the then British Broadcasting Company (Mr J.C.W. Reith) and the Wheatstone Professor of Physics at King's College (Professor Appleton). Afterwards the party went out into the rain and Sir Charles Sherrington, President of the Royal Society, unveiled the plaque. In his speech he said that though Wheatstone received many honours in his lifetime 'the far-reaching character of his achievements could not have been so evident as they are now, viewed along the perspective of the years'.[1] The plaque at Gloucester has been removed but in 1998 a plaque was unveiled on the shop at 52/54 Westgate Street which had been his father's shop and where he lived as a child. Earlier, in 1975, a plaque was unveiled on the house in Park Crescent, London.[2] Wheatstone now has the unusual distinction of two Blue Plaques to his memory. Both unveiling ceremonies were accompanied by music on the concertina.

How should we assess him after more than 100 years? W.H. Preece, later Sir William Preece, Engineer-in-Chief of the Post Office, said of Wheatstone that he was not a philosopher nor a 'deep investigator', but essentially a designer of delicate apparatus.[3] Preece, who was born in 1834, was speaking from his personal knowledge of Wheatstone and his assessment agrees with all we know of Wheatstone in later life. As a young man Wheatstone *had* been a 'deep investigator', but the direction of his work had changed over the years. He began by studying the working of the musical instruments among which he was brought up and investigated the physical principles of sounding bodies. Before long he was applying his scientific discoveries to the design of new musical instruments and to his dream of communicating over a distance.

Another characteristic noted by Preece was that:

His bibliographical knowledge was almost incredible. He seemed to know every book that was written and every fact recorded, and anyone in doubt had only to go to Wheatstone to get what he wanted.

The contribution Wheatstone made to the rapidly expanding sciences of his day by his knowledge of the literature and his wide personal contacts at home and abroad is hard to evaluate but the comments of many of those associated with him testify to its importance.

Wheatstone's greatest work was in electricity and the electric telegraph, yet he discovered hardly any new facts in electrical science. His vital contribution here was to study and apply the researches of others; when this required the design and construction of some new and delicate apparatus, Wheatstone excelled.

His musical instruments have been largely forgotten, although the concertina enjoyed a revival in the 1960s and 1970s, but the credit for the discoveries in acoustics will always be his. When stereoscopic pictures are viewed today, it is by means Wheatstone did not know, but he discovered the fundamental principle. For many purposes the telegraph soon gave way to the telephone. When the first edition of this book was written in 1974, I said Wheatstone would have preferred to see the telex – an advanced printing telegraph – in every home and office, rather than the telephone with its bell demanding immediate attention. Since then the telex has been superseded by the fax and now the e-mail. Wheatstone would have delighted in these further developments from his telecommunications work. The e-mail is surely a logical development of the printing telegraph. The person sending a message types characters which in due course are printed on the recipient's machine.

With the benefit of hindsight we can see that his electric motors were impractical, being 'magnetic' rather than 'electromagnetic' machines; with the development of power semiconductors reluctance motors have now become an attractive option for some applications. Even if Wheatstone's motor researches took him at that time into a blind alley, it was a blind alley that needed to be explored. At that pioneering stage no one could know which lines of research would bear fruit. The patient investigation of each possibility by Wheatstone and the other pioneers of his generation prepared the way for the founders of the electrical industry in the last quarter of the century. Wheatstone was one of

Figure 16.2   Painting of Wheatstone in the Institution of Electrical Engineers.

Source: By courtesy of the IEE Archivist.

the independent inventors of the self-excited generator, one of the most important developments leading to the electric power industry, but he died a little too soon to see the beginning of public electricity supply. If he had lived another ten years, his house would surely have shone with the electric light as well as with the wonders of the telegraph.

In electrical measurements some of Wheatstone's methods are still in use when resistances are put in parallel with or in series with a meter to adjust its range. Most of his other work has been superseded, but the applied scientist must expect to be overtaken by those who come after. Wheatstone was confident that these subjects were worth pursuing whether or not it was he who finally brought about their practical realization.

His true memorials are the laboratory and Chair at King's College, the tradition of research that goes with them, and the corpus of knowledge he helped to gather.

Even the Wheatstone Bridge is being overtaken by modern electronics but it remains a fitting memorial, not because he devised it – clearly he did not – but because it epitomizes Wheatstone at his best: seeing a neglected idea in the literature of science, experimenting with it, and applying it to practical ends.

Notes

1 There is an account of the proceedings in Gloucester on the 50th anniversary of Wheatstone's death in Gloucester City Library, Local History Collection, and *The Gloucester Journal* 10 and 24 October 1925.
2 The writer was present on both occasions.
3 W.H. Preece, 'The Telegraphic Achievements of Wheatstone', *Proceedings of the Royal Institution*, 13 February 1880, pp. 297–304.

Appendix – sources

There is no single collection of Wheatstone papers, and the sources for this book are found in numerous different places.

Wheatstone's correspondence

Correspondence with Cooke, and papers relating to the telegraph arbitration, are in the Archives of the Institution of Electrical Engineers.

Also in the IEE are correspondence with Ronalds and some correspondence with Faraday.

Most of the Faraday correspondence is in the Royal Institution. All is transcribed and published in Frank A.J.L. James, *The Correspondence of Michael Faraday*, vol. 1, 1991 and subsequent volumes.

Correspondence with Herschel and Roget is in the Royal Society Archives.

Correspondence with Dumas is in the Académie des Sciences in Paris.

Correspondence with the General Board of Health and with the Commissioners of Works is in King's College London, which also has such other papers of Wheatstone's as survive.

Records of the Wheatstone musical business are in the Concertina Museum collections.

Wheatstone's scientific papers

Most, but not all, of Wheatstone's scientific papers were reprinted after his death in a collected volume published by the Physical Society of London and edited by Professor W.G. Adams, Dr Atkinson, Latimer Clark, Professor G.C. Foster and Robert Sabine. This list includes a number of papers not found either in the collected volume or in the Royal Society Catalogue of Scientific Papers. All these papers were published in Wheatstone's name, except where it is indicated that a paper was published anonymously or over initials only.

The papers are divided by subject thus:

> Papers relating to sound
> Papers relating to electricity and the telegraph
> Papers relating to optics
> Miscellaneous papers
> Patents

The following abbreviations are used for journals:

Annal de Chimie	*Annales de Chimie et de Physique*
BA Report	*Report of the British Association for the Advancement of Science*
Bibl. Univ.	Bibliothéque Universelle de Genéve
Comptes Rendus	*Comptes Rendus de l'Académie des Sciences*
Froriep Notizen	(Froriep's) *Notizen aus dem Gebiete der Natur- und Heilkunde*
JRI	*Journal of the Royal Institution*
Majocchi, Ann Fis Chim	(Majocchi's) *Annali di Fisica, chimica e matematiche*
Phil Mag	*The Philosophical Magazine*
Phil Trans	*Philosophical Transactions of the Royal Society of London*
QJS	*Quarterly Journal of Science, Literature and Art*
Schweigger, Journ	*Journal für Chemie und Physik ... vom* Dr. J.S.C. Schweigger

Papers relating to sound

New Experiments on Sound
  (Thomson's) Annals of Philosophy, **6**, 1823, pp. 81–90
  Annal de Chimie, B23B, 1823, pp. 313–22
  Schweigger's Journ., **42**, 1824, pp. 185–201

Explanation of the Harmonic Diagram
  An Explanation of the Harmonic Diagram invented by C. Wheatstone, published by C. Wheatstone, 436 The Strand, London.

Description of the Kaleidophone, or Phonic Kaleidoscope; a new Philosophical Toy, for the Illustration of several Interesting and Amusing Acoustical and Optical Phenomena.
  QJS, second quarter 1827, pp. 344–51
  Poggend. Annal, **10**, 1827, pp. 470–80

Experiments on Audition
  QJS, third quarter 1827, pp. 67–72
  Froriep, Notizen, **19**, 1828, col. 81–5

On the Resonances, or Reciprocated Vibrations of Columns of Air
QJS, first quarter 1828, pp. 175–83
Schweigger, Journ, **53**, 1828, pp. 327–33

On the Transmission of Musical Sounds through Solid Linear Conductors, and on their subsequent Reciprocation
JRI, **2**, 1831
Froriep, Notizen, **27**, 1830, col. 129–38
Poggend, Annal, **26**, 1832, pp. 251–68

On the Vibrations of Columns of Air in Cylindrical and Conical Tubes
Athenaeum, 24 March 1832, p. 194

On the Figures obtained by strewing Sand on Vibrating Surfaces commonly called Acoustic Figures
Phil Trans, 1833, pp. 593–634

On the various attempts which have been made to imitate human speech by mechanical means
BA Report, 1835, p. 14

Experimental Verification of Bernouilli's Theory of Wind Instruments
BA Report, 1835, p. 558

A review article published over the initials 'C.W.' about:
1. On the Vowel Sounds, and on Reed Organ-Pipes, by Robert Willis,
2. Le Méchanisme de la Parole, suivi de la Description d'une machine parlante, by M. de Kempelen,
3. C.G. Kratzenstein, Tentamen Coronatum de voce,
   in London and Westminster Review, Nos. xi and liv, October 1837

Papers relating to electricity and the telegraph

An Account of some Experiments to measure the Velocity of Electricity and the Duration of Electric Light
Phil Trans, 1834
Poggend, Annal, **34**, 1835, pp. 464–79
Archives de l'Electricité, par A. de la Rive, **2**, 1842, pp. 37–53

On the Thermo-electric Spark
Phil Mag, **10**, 1837, pp. 414–17
Poggend, Annal., 41, 1837, pp. 160–63

Description of the Electro-magnetic Clock
Proc Roy Soc, **4**, 1840, pp. 249–50

Account of an Electro-magnetic Telegraph
(Sturgeon) Annals of Electricity, **5**, 1840, pp. 337–49

Letter to Col. Sabine on a new Meteorological Instrument
  BA Report 1842, pt. 2, p. 9

Description of the Telegraph Thermometer
  BA Report 1843, pp. 128–9

An account of several new Instruments and Processes for determining the Constants of a Voltaic Circuit. The Bakerian Lecture for 1843.
  Phil Trans, **133**, 1843, pp. 303–27
  Annal de Chimie, **10**, 1844, pp. 257–98
  Archives de l'Electricite', **4**, 1844, pp. 102–43
  Majocchi, Ann Fis Chim, **15**, 184, pp. 148, 164, 250–73
  Poggend, Annal., **42**, 184, pp. 499–543

Enregistreur électromagnétique pour les Observations Météorologiques
  Archives de l'Electricité, **4**, 1844, pp. 70–72
  Majocchi, Ann Fis Chim, **14**, 1844, pp. 244–7

Note sur le Chronoscope électromagnétique
  Comptes Rendus, **20**, 1845, pp. 1554–61
  Poggend, Annal., **45**, 1845, pp. 451–60
  (Walker) Electrical Magazine, **2**, 1846, pp. 86–93

An Account of some Experiments made with the Submarine Cable of the Mediterranean Electric Telegraph
  Proc Roy Soc, **7**, 1855, pp. 328–33
  Annal de Chimie, **46**, 1856, pp. 121–4
  Poggend, Annal., **96**, 1855, pp. 164–71

On the position of Aluminium in the Voltaic Series
  Proc Roy Soc, **7**, 1855, pp. 369–70

Télégraphe automatique écrivant
  Comptes Rendus, **48**, 1859, pp. 214–20

(Jointly with F.A. Abel)
Report to the Secretary of State for War on The Application of Electricity from Different Sources to the Explosion of Gunpowder, November 1860
  HMSO 1861

On the Circumstances which influence the Inductive Discharges of Submarine Telegraphic Cables. Report of the Joint Committee appointed by the Lords of the Committee of Privy Council for Trade and the Atlantic Telegraph Company, to inquire into the construction of Submarine Telegraph Cables,
  HMSO 1861

On a New Telegraphic Thermometer, and on the Application of the Principle of its construction to other Meteorological Indicators
  BA Report, 1867, pp. 11–13

On the Augmentation of the Power of a Magnet by the reaction thereon of Currents induced by the Magnet itself
  Proc Roy Soc, **15**, 1867, pp. 369–72

On a cause of Error in Electroscopic Experiments
  Proc Roy Soc, **18**, 1870, pp. 330–33

Papers relating to optics

(Published anonymously)
  Contributions to the Physiology of Vision. No. 1
  JRI, **1**, October 1830, pp. 101–17

(Published over the initials 'C.W.')
  Contributions to the Physiology of Vision. No. 11
  JRI, **1**, May 1831, pp. 534–7

Remarks on Purkinje's Experiments
  BA Report, **1835**, pp. 551–3

Contributions to the Physiology of Vision-Part the First. On some remarkable, and hitherto unobserved, Phenomena of Binocular Vision
  Phil Trans, 1838
  Annal de Chimie, **2**, 1841, pp. 330–70
  Poggend, Annal., Supplementary volume 1, 1842, pp. 1–48

On binocular vision; and on the stereoscope, an instrument for illustrating its phenomena
  BA Report, 1838, pt. 2, pp. 16–17
  Bibl. Univ., **17**, 1838, pp. 174–5
  Froriep, Notizen, **12**, 1839, col. 176–7
  Poggend, Annal., **47**, 1839, pp. 625–7

On a singular Effect of the Juxtaposition of certain Colours under particular circumstances
  BA Report, 1844, p. 10

On a means of determining the apparent Solar Time by the Diurnal Changes of the Plane of Polarization at the North Pole of the Sky
  BA Report, 1848, pp. 10–12

Contributions to the Physiology of Vision-Part the Second. On some remarkable, and hitherto unobserved, Phenomena of BinocularVision
  Phil Trans, 1852
  Phil Mag, **3**, 1852, pp. 241–67, 504–23

On the Binocular Microscope, and on stereoscopic pictures of microscopic objects
  Transactions of the Microscopical Society, **I**, 1853, pp. 99–102

Experiments on the Successive Polarization of Light, with the Description of a new Polarizing Apparatus
   Proc Roy Soc, **19**, 1871, pp. 381–9

Miscellaneous papers

On the Prismatic Decomposition of Electrical Light
   BA Report 1835, pp. 11–12
   Poggend, Annal., **36**, 1835, pp. 148–9

(Jointly with T. R. Robinson)
Report of the Committee for conducting experiments with captive balloons
   BA Report, 1843, pp. 128–9

Note relating to M. Foucault's new Mechanical Proof of the Rotation of the Earth
   Proc Roy Soc, **6**, 1851, pp. 65–8
   Poggend, Annal., **83**, 1851, pp. 306–8

On Fessel's Gyroscope
   Proc Roy Soc, **7**, 1854, pp. 43–8

On the Formation of Powers from Arithmetical Progressions
   Proc Roy Soc, **7**, 1855, pp. 145–51

Interpretation of an important Historical Document in Cipher
   Memoirs of the Philobiblion Society, 1862

Instructions for the Employment of Wheatstone's Cryptograph
   Pamphlet published to accompany an instrument called 'The Cryptograph'

Wheatstone's Patents

Number	Date	Subject

England (up to 1852)

5803	19 June 1829	Wind musical instruments
7154	27 July 1836	Wind musical instruments
7390	12 June 1837	Electric telegraphs

(Obtained jointly with W.F. Cooke. There is a corresponding Scottish patent dated 12 December 1837 and an Irish patent dated 23 April 1838)

| 8345 | 21 January 1840 | Electric telegraphs |

(Obtained jointly with W.F. Cooke. There is a corresponding Scottish patent dated 21 August 1840 and an Irish patent dated 27 October 1840)

| 9022 | 7 July 1841 | Producing, regulating and applying electric currents |

10041	8 February 1844	Concertina and other musical instruments in which the sounds are produced by the action of wind on vibratory springs
10655	6 May 1845	Electric telegraphs and other apparatus

(Obtained jointly with W.F. Cooke. There is a corresponding Scottish patent dated 3 July 1845 and an Irish patent dated 22 October 1845)

United Kingdom (1852 onwards)

1239/1858	2 June 1858	Electric telegraphs, and apparatus connected therewith
1241/1858	2 June 1858	Electro-magnetic telegraphs and apparatus
2462/1860	10 October 1860	Electro-magnetic telegraphs
220/1867	28 January 1867	Electric telegraphs
2897/1870	3 November 1870	Electric telegraphs

(Obtained jointly with J.M.A. Stroh)

2172/1871	18 August 1871	Electric telegraphs; magneto-electric machines

(Obtained jointly with J.M.A. Stroh)

39/1872	4 January 1872	Musical instruments

(Obtained jointly with J.M.A. Stroh)

473/1872	15 February 1872	Electric telegraphs; magnetoelectric machines

(Obtained jointly with J.M.A. Stroh)

2771/1875	5 August 1875	Electric telegraphs

(Patent granted to R. Sabine as executor of Sir C. Wheatstone)

Index

Figures in italics indicate pages with illustrations.
'f' following a page number indicates that the relevant term continues on the next page.
'ff' following a page number indicates that the relevant term continues over several pages.

Printed in the USA
CPSIA information can be obtained
at www.ICGtesting.com
JSHW011509221024
72173JS00005B/1256

9 780852 961032